Manufacturing of Fibrous Composites for Engineering Applications

Manufacturing of Fibrous Composites for Engineering Applications

Editor

Jinyang Xu

MDPI • Basel • Beijing • Wuhan • Barcelona • Belgrade • Manchester • Tokyo • Cluj • Tianjin

MDPI

Editor
Jinyang Xu
Shanghai Jiao Tong University
Shanghai, China

Editorial Office
MDPI
St. Alban-Anlage 66
4052 Basel, Switzerland

This is a reprint of articles from the Special Issue published online in the open access journal *Journal of Composites Science* (ISSN 2504-477X) (available at: https://www.mdpi.com/journal/jcs/special_issues/manufacturing_fibrous_composites_engineering_applications).

For citation purposes, cite each article independently as indicated on the article page online and as indicated below:

LastName, A.A.; LastName, B.B.; LastName, C.C. Article Title. *Journal Name* **Year**, *Volume Number*, Page Range.

ISBN 978-3-0365-8120-0 (Hbk)
ISBN 978-3-0365-8121-7 (PDF)

Contents

About the Editor

Jinyang Xu

Jinyang Xu is an Associate Professor and a Doctoral Supervisor of Mechanical Engineering at Shanghai Jiao Tong University, China. He received his Ph.D. (2016) in Mechanical Engineering from Arts et Métiers ParisTech, France. His research interests focus on composites machining, numerical modeling, micro/nano cutting, and surface texturing. He has published over 100 articles in highly-ranked JCR-referenced journals as a first/corresponding author, and edited 2 monographs, 5 book chapters, 6 int. conf. proceedings & 8 special issues. He now serves as the Editor-in-Chief of Journal of Coating Science and Technology (JCST), International Journal of Precision Technology (IJPTech), and International Journal of Product Sound Quality (IJPSQ). He is an Associate Editor of Proc. Inst. Mech. Eng. Pt. E - J. Process. Mech. Eng. (SCIE/EI), Simulation - Transactions of the Society for Modeling and Simulation International (SCIE/EI), and Frontiers in Materials (SCIE/EI). Presently, he is an Editorial Board Member of 7 SCI journals, including Green Materials, Journal of Superhard Materials, International Journal of Aerospace Engineering, Coatings, Lubricants, etc. He was honored with the prestigious IAAM Scientist Medal of the year 2020, and awarded the Shanghai Pujiang Scholar by the Shanghai Municipality in 2017. He is a fellow of IAAM and also a senior member of ASME and SCS.

Journal of
Composites Science

Editorial

Manufacturing of Fibrous Composites for Engineering Applications

Jinyang Xu

State Key Laboratory of Mechanical System and Vibration, School of Mechanical Engineering, Shanghai Jiao Tong University, Shanghai 200240, China; xujinyang@sjtu.edu.cn

Fibrous composites are advanced engineering materials featuring the impregnation of fiber phase with a polymer matrix base to yield enhanced properties. They can be categorized in terms of the fiber type, such as aramid-fiber-reinforced polymer (AFRP), carbon-fiber-reinforced polymer (CFRP), glass-fiber-reinforced polymer (GFRP), Kevlar-fiber-reinforced polymer (KFRP), etc. They can also be classified into continuous and discontinuous fiber composites. In general, discontinuous fiber composites show better isotropic behavior than continuous fiber composites, as short fibers can enhance the matrix in every direction [1]. The fibrous composites are often fabricated with a polymer matrix, which can be either thermoset polymers or thermoplastic polymers. The thermoset polymers have the property of becoming permanently hard and rigid when heated or cured. The thermoplastic polymers entail the linear chain of the molecular structures; thus, they can be recycled and reused when heated [2–4].

Since fibrous composites show higher specific mechanical and physical properties than conventional alloys and steels [1,5], they are very attractive in modern aerospace industries [6–9]. For instance, CFRP composites are often employed in the aircraft wing boxes, horizontal and vertical stabilizers, and wing panels [10]. GFRP composites are mainly used in the fairings, storage room doors, landing gear doors, and passenger compartments [10,11]. Reports indicate that most international aircraft manufacturers, including Airbus and Boeing, are seeking the use of fibrous composites to fabricate large load-carrying structures favoring the energy saving of new-generation airplanes. Despite the net shapes of molded fibrous composites, secondary manufacturing operations are still required to create the target shapes and desired quality [8,9,12,13]. However, fibrous composites are extremely difficult to machine, as they possess rather poor machinability compared to conventional alloys and steels. The term "machinability" signifies the degree of difficulty of cutting a workpiece material with qualified quality. The poor machinability of fibrous composites may arise from the inherent anisotropic behaviors of the fiber/matrix system and the heterogeneous architecture of the composites. The most-used machining operations for shaping fibrous composites mainly include trimming, turning, milling, and drilling. Since the reinforcing fibers and the matrix base show completely different properties, machining of these composites has posed huge challenges to the current manufacturing community. The specific issues associated with the cutting of fibrous composites are difficult chip removal, poor surface quality, and rapid tool wear.

Chip separation mode is a critical procedure determining the eventual cutting responses of workpiece materials. However, the cutting mechanisms of fibrous composites are much complicated than those of conventional metallic alloys, as the composite separation modes are fiber-orientation dependent. Many studies have attributed the fiber-orientation-dependent chip removal to the effects of the fiber layup on the properties of chips under specific cutting loads. Consequently, producing consistent surface finish of cut composites is very challenging. The fundamental chip removal modes encountered in composite machining mainly include shear-induced fracture, bending-induced fracture, and fiber/matrix interfacial debonding, depending significantly on the varying fiber cutting

Citation: Xu, J. Manufacturing of Fibrous Composites for Engineering Applications. *J. Compos. Sci.* **2022**, *6*, 187. https://doi.org/10.3390/jcs6070187

Received: 22 June 2022
Accepted: 23 June 2022
Published: 24 June 2022

angle [14–16]. Additionally, cutting forces and machining temperatures are also firmly related to the variation of the fiber cutting angle during the machining of fibrous composites. Normally, lower magnitudes of cutting forces/temperatures can be produced under the along fiber cutting relation, in which the fiber cutting angle is acute. In contrast, chip separation occurring under the against fiber cutting relation is often considered unfavorable for producing desired surface quality for fibrous composites. Therefore, selecting a proper fiber-cutting angle plays a crucial role in the high-quality machining of fibrous composites

Poor surface quality is another critical issue associated with fibrous composite machining. Since the removal mechanisms of fibers and matrix are different and change continuously with the fiber layup, even under the action of the identical cutting edges, serious defects involving delamination, burrs, tearing, surface cavities, and glass transition failure easily occur, which can greatly deteriorate the quality of machined composite parts [17]. Additionally, the aforementioned damage not only reduces the surface finish and assembly tolerance, but also affects the fatigue strength of cut holes, accounting for a large proportion of part rejections [18–21]. Moreover, delamination and tearing are often noted as the most critical failures of fibrous composites, and should be carefully suppressed during the machining operations as they cannot be repaired once they occur. Functionally designed special tools are feasible for minimizing the generation of various cutting-induced damages for fibrous composites. The use of optimized cutting parameters can also benefit the improvement of the surface quality of cut fibrous composites. To deal with cutting-induced damages, future endeavours must optimize the process parameters, develop functional cutting tools and use advanced machining techniques for fibrous composites.

Rapid tool wear is a crucial issue when cutting fibrous composites, as hard fibers induce severe abrasions/modifications onto the tool surfaces and lead to the blunting of tool edges. The primary wear mode frequently encountered in the machining of fibrous composites is abrasion wear in the form of cutting-edge rounding (CER) [22–25]. Additionally, CER has become a significant indicator for the assessment of tool wear, particularly for drilling fibrous composites, which provides a more accurate quantification than the conventionally used wear-width indicator [1]. As the composite chips are often separated within a small tool–chip interface, crater wear shows no ability to take place. Progressive abrasion wear often results in undesired consequences when machining fibrous composites, such as increased cutting forces, excessive machining temperatures, deteriorated surface quality, and reduced tool life. Moreover, the dominant tool failure in machining fibrous composites is edge chipping or coating peeling during the chip removal process. Developing superior tool coating materials and optimizing tool geometries would be a feasible solution to the tool wear issues for fibrous composites.

In the end, with the rapid development of manufacturing technologies, various types of advanced machining methods have been developed in recent years, providing new prospects for achieving damage-free machining of fibrous composites. For instance, helical milling, variable feed drilling, and ultrasonic vibration-assisted machining have emerged and have been extensively used in the machining of fibrous composites, which yield outstanding performances in suppressing cutting-induced damages. Moreover, intelligent manufacturing technology, which has become a hot research focus in both academia and industry, can be applied to composites machining to realize the in-situ detection and high-precision control of online machining status. Then, high-efficiency and high-quality machining of fibrous composites can be accomplished.

Funding: The work was supported by the National Natural Science Foundation of China (Grant No. 52175425) and the 9th Sino-Hungarian Intergovernmental Scientific and Technological Cooperation Project (Grant No. 2021-07).

Conflicts of Interest: The author declares no conflict of interest.

References

1. Xu, J.; Davim, J.P. Machining of Fibrous Composites: Recent Advances and Future Perspectives. In *Mechanical and Industrial Engineering: Historical Aspects and Future Directions*; Davim, J.P., Ed.; Springer International Publishing: Cham, Switzerland, 2022; pp. 161–177.
2. Ahmad, F.; Manral, A.; Bajpai, P.K. Machining of thermoplastic composites. In *Processing of Green Composites*; Rakesh, P.K., Singh, I., Eds.; Springer: Singapore, 2019; pp. 107–123.
3. Xu, J.; Huang, X.; Davim, J.P.; Ji, M.; Chen, M. On the machining behavior of carbon fiber reinforced polyimide and PEEK thermoplastic composites. *Polym. Compos.* **2020**, *41*, 3649–3663. [CrossRef]
4. Xu, J.; Huang, X.; Chen, M.; Davim, J.P. Drilling characteristics of carbon/epoxy and carbon/polyimide composites. *Mater. Manuf. Process.* **2020**, *35*, 1732–1740. [CrossRef]
5. Xu, J.; Davim, J.P.; Chen, M. Machining Effects of Fibrous Composites and Related Stacks for Aerospace Applications. In *Advanced Composites in Aerospace Engineering Applications*; Mazlan, N., Sapuan, S.M., Ilyas, R.A., Eds.; Springer International Publishing: Cham, Switzerland, 2022; pp. 109–125.
6. Abrão, A.M.; Faria, P.E.; Rubio, J.C.C.; Reis, P.; Davim, J.P. Drilling of fiber reinforced plastics: A review. *J. Mater. Process. Technol.* **2007**, *186*, 1–7. [CrossRef]
7. Geier, N.; Davim, J.P.; Szalay, T. Advanced cutting tools and technologies for drilling carbon fibre reinforced polymer (CFRP) composites: A review. *Compos. Part A Appl. Sci. Manuf.* **2019**, *125*, 105552. [CrossRef]
8. Liu, D.; Tang, Y.; Cong, W.L. A review of mechanical drilling for composite laminates. *Compos. Struct.* **2012**, *94*, 1265–1279. [CrossRef]
9. Aamir, M.; Tolouei-Rad, M.; Giasin, K.; Nosrati, A. Recent advances in drilling of carbon fiber–reinforced polymers for aerospace applications: A review. *Int. J. Adv. Manuf. Technol.* **2019**, *105*, 2289–2308. [CrossRef]
10. Soutis, C. Fibre reinforced composites in aircraft construction. *Prog. Aerosp. Sci.* **2005**, *41*, 143–151. [CrossRef]
11. Gay, D.; Hoa, S.V.; Tsai, S.W. *Composite Materials Design and Applications*; CRC Press: New York, NY, USA, 2003.
12. Xu, J.; Lin, T.; Davim, J.P. On the machining temperature and hole quality of CFRP laminates when using diamond-coated special drills. *J. Compos. Sci.* **2022**, *6*, 45. [CrossRef]
13. Che, D.; Saxena, I.; Han, P.; Guo, P.; Ehmann, K.F. Machining of carbon fiber reinforced plastics/polymers: A literature review. *J. Manuf. Sci. Eng. Trans. ASME* **2014**, *136*, 034001. [CrossRef]
14. Wang, D.H.; Ramulu, M.; Arola, D. Orthogonal cutting mechanisms of graphite/epoxy composite. Part I: Unidirectional laminate. *Int. J. Mach. Tools Manuf.* **1995**, *35*, 1623–1638. [CrossRef]
15. Wang, D.H.; Ramulu, M.; Arola, D. Orthogonal cutting mechanisms of graphite/epoxy composite. Part II: Multi-directional laminate. *Int. J. Mach. Tools Manuf.* **1995**, *35*, 1639–1648. [CrossRef]
16. Iliescu, D.; Gehin, D.; Iordanoff, I.; Girot, F.; Gutiérrez, M.E. A discrete element method for the simulation of CFRP cutting. *Compos. Sci. Technol.* **2010**, *70*, 73–80. [CrossRef]
17. Xu, J.; Yin, Y.; Davim, J.P.; Li, L.; Ji, M.; Geier, N.; Chen, M. A critical review addressing drilling-induced damage of CFRP composites. *Compos. Struct.* **2022**, *294*, 115594. [CrossRef]
18. Geng, D.; Zhang, D.; Li, Z.; Liu, D. Feasibility study of ultrasonic elliptical vibration-assisted reaming of carbon fiber reinforced plastics/titanium alloy stacks. *Ultrasonics* **2017**, *75*, 80–90. [CrossRef]
19. Fleischer, J.; Teti, R.; Lanza, G.; Mativenga, P.; Moehring, H.-C.; Caggiano, A. Composite materials parts manufacturing. *CIRP Ann. Manuf. Technol.* **2018**, *67*, 603–626. [CrossRef]
20. Wang, G.-D.; Melly, S.K.; Li, N. Using dampers to mitigate thrust forces during carbon-fibre reinforced polymer drilling: Experimental and finite element evaluation. *J. Reinf. Plast. Compos.* **2018**, *37*, 60–74. [CrossRef]
21. Al-Wandi, S.; Ding, S.; Mo, J. An approach to evaluate delamination factor when drilling carbon fiber-reinforced plastics using different drill geometries: Experiment and finite element study. *Int. J. Adv. Manuf. Technol.* **2017**, *93*, 4043–4061. [CrossRef]
22. Faraz, A.; Biermann, D.; Weinert, K. Cutting edge rounding: An innovative tool wear criterion in drilling CFRP composite laminates. *Int. J. Mach. Tools Manuf.* **2009**, *49*, 1185–1196. [CrossRef]
23. Wang, X.; Kwon, P.Y.; Sturtevant, C.; Kim, D.; Lantrip, J. Tool wear of coated drills in drilling CFRP. *J. Manuf. Process.* **2013**, *15*, 127–135. [CrossRef]
24. Çelik, A.; Lazoglu, I.; Kara, A.; Kara, F. Wear on SiAlON ceramic tools in drilling of aerospace grade CFRP composites. *Wear* **2015**, *338*, 11–21. [CrossRef]
25. Xu, J.; Lin, T.; Davim, J.P.; Chen, M.; El Mansori, M. Wear behavior of special tools in the drilling of CFRP composite laminates. *Wear* **2021**, *476*, 203738. [CrossRef]

Journal of
Composites Science

Review

A Review on the Fabrication and Mechanical Characterization of Fibrous Composites for Engineering Applications

H. S. Ashrith [1,*], T. P. Jeevan [1] and Jinyang Xu [2,*]

1 Department of Mechanical Engineering, Malnad College of Engineering, Hassan 573202, India; jeevantpmce@gmail.com
2 State Key Laboratory of Mechanical System and Vibration, School of Mechanical Engineering, Shanghai Jiao Tong University, Shanghai 200240, China
* Correspondence: ashuhiriyal@gmail.com (H.S.A.); xujinyang@sjtu.edu.cn (J.X.)

Abstract: This review focuses on the fabrication and mechanical characterization of fibrous composites for engineering applications. Fibrous composites are materials composed of two or more distinct phases, with fibers embedded in a matrix. The properties of these materials depend on the properties of both the fibers and the matrix, as well as the way they are combined and fabricated. The various fabrication methods, along with the process parameters, used to manufacture synthetic and natural fibrous composites for engineering applications, including hand lay-up, compression molding, resin transfer molding, additive manufacturing, etc., are discussed. The mechanical characterization of fibrous composites, including their strength, stiffness, and toughness of both synthetic and natural fibrous composites are discussed. The advantages and disadvantages of fiber reinforcement are discussed, along with their influence on the resulting mechanical characteristics of the composites. It can be observed that the mechanical properties of fibrous composites can be tailored by controlling various factors, such as the fiber orientation, fiber volume fraction, and matrix type. Although fibrous composites offer significant advantages, several challenges hinder their widespread use in engineering applications. These challenges include high manufacturing costs, limited design guidelines, and difficulties in predicting their mechanical behavior under various loading conditions. Therefore despite their unique properties, these challenges must be overcome for fibrous composites to realize their full potential as high-performance materials.

Keywords: FRPC; manufacturing techniques; additive manufacturing; mechanical characterization

Citation: Ashrith, H.S.; Jeevan, T.P.; Xu, J. A Review on the Fabrication and Mechanical Characterization of Fibrous Composites for Engineering Applications. *J. Compos. Sci.* **2023**, *7*, 252. https://doi.org/10.3390/jcs7060252

Academic Editor: Francesco Tornabene

Received: 25 May 2023
Revised: 6 June 2023
Accepted: 15 June 2023
Published: 18 June 2023

1. Introduction

Fibrous composites are a highly valued group of materials due to their exceptional mechanical characteristics, lightness, and versatility in various engineering applications These materials possess anisotropic qualities, meaning that their properties vary based on the direction of loading. Fibrous composites are fabricated by combining two or more materials, which include matrix and reinforcing fibers, which possess varying mechanical properties. The matrix material serves to connect the fibers and transfer loads between them, while the fibers offer strength and stiffness to the composite. Various materials, such as glass, carbon, aramid, and natural fibers, can be used as reinforcements in composite fabrication [1–5]. Fibrous composites offer a significant advantage in terms of their high strength-to-weight ratio, making them highly desirable for applications where weight reduction is very crucial. The aerospace industry, for instance, has extensively utilized carbon fiber reinforced polymer (CFRP) composites for their exceptional strength and stiffness properties [6,7]. According to Khan et al. [8] using CFRP composites in aircraft wings resulted in a 15% decrease in weight and a 30% improvement in fuel efficiency in comparison to conventional metallic materials. Besides the aerospace sector, the automotive industry has also made extensive use of fibrous composites to reduce weight and enhance fuel efficiency. The lightweight and high-strength properties of fibrous composites make them appealing for use in sporting goods applications. The use of CFRP composites in bicycle frames resulted in a 30% weight reduction and a 20% increase in stiffness compared to traditional aluminum frames [9]. Kim et al. [10] examined the application of glass fiber reinforced polymer (GFRP) composites for automotive door inner panels and concluded that these composites resulted in a 36% reduction in weight relative to traditional steel panels, without any compromise in mechanical properties. Stepanova and Korzhikova-Vlakh [11] investigated the use of cellulose nanofibers to reinforce polylactic acid (PLA) composites and found that the resulting composites exhibited improved tensile strength and modulus. Researchers are currently exploring new methods for enhancing the properties of fibrous composites.

Despite their benefits, the cost of fibrous composites remains high, limiting their use in some applications. Furthermore, these materials can be damaged by impact or other forms of loading, which can decrease their reliability and durability over time. Fibrous composites can be fabricated by employing different techniques, such as hand lay-up (HL), filament winding (FW), pultrusion, resin transfer molding (RTM), and additive manufacturing (AM), each having their own benefits and drawbacks. The choice of technique depends on the desired properties of the composite material. Various investigations have been performed on the fabrication of fibrous composites to enhance their mechanical properties and reduce the cost of production. Some studies have explored the usage of flax and jute fibers as a cost-effective and sustainable alternative to synthetic fibers [12,13], while others have looked at improving the bonding between the matrix and fibers to enhance the mechanical performance of the composite material [14].

The mechanical characterization of fibrous composites is critical for evaluating their performance and potential use in engineering applications. Mechanical characterization involves the measurement and analysis of the properties, such as strength, stiffness, toughness, and fatigue resistance [15,16]. Understanding this behavior is necessary to access and optimize the performances of fibrous composites under different loading conditions. Various techniques are used to evaluate the mechanical performance of fibrous composites, including tensile, compressive, shear, and bending tests. These tests are usually performed under varying loads to determine the mechanical characteristics of composites in different directions. Additionally, non-destructive assessment techniques, for example ultrasonic testing, X-ray tomography, and thermography, have also been developed to assess the defects and internal structure of composites. Extensive research has been carried out to address the challenges for improving the performance of fibrous composites and their applications. While several detailed reviews are available on CFRP composites, GFRP composites, or natural fiber-reinforced polymer (NFRP) composites, no single review covers all types of fibrous composites and their detailed fabrication processes. Therefore, this paper presents a comprehensive review of the manufacturing techniques of fibrous composites, including a mechanical characterization and factors influencing mechanical properties.

2. Fabrication Methods of Fibrous Composites

The fabrication of fiber-reinforced polymer composites (FRPCs) is a complex phenomenon and includes several techniques. FRPC fabrication techniques can be broadly classified into conventional, automated, and additive manufacturing techniques. The fabrication technique employed primarily depends on the size and application of the composites. Among the available techniques, conventional technologies, such as hand lay-up, injection molding, compression molding, resin transfer molding, and filament winding, are widely used to fabricate the FRPC [1]. This section offers a brief description of the commonly used techniques in fabricating fibrous composites made of synthetic and natural fibers for engineering applications.

Hand lay-up is the commonly employed technique in fabricating FRPCs. It involves laying the fiber reinforcements on the mold followed by pouring a thermosetting resin to wet the reinforcements. The resin is spread uniformly on the laid reinforcements using a brush or a roller followed by laying the next layer and repeating the process until the desired thickness is obtained. Normally, a silicone demolding agent is applied to the mold surface before laying reinforcements to facilitate easy removal of the composites after curing [17]. Figure 1a illustrates the HL technique employed for composite fabrication.

The compression molding (CM) technique is employed for sheet molding compounds (SMCs) and bulk molding compounds (BMCs). SMCs and BMCs are semi-cured sheets of thermoset resin reinforced with short fibers or impregnated fibers, inert fillers, catalysts, pigments and stabilizers, release agents, and thickeners. BMCs are stored at a lower temperature to avoid hardening before subjecting them to compression molding [18] Molding compounds are put in a hot mold and cured by pressing (Figure 1b). The pressure and temperature applied during the process depend on the material type and desired thickness [1].

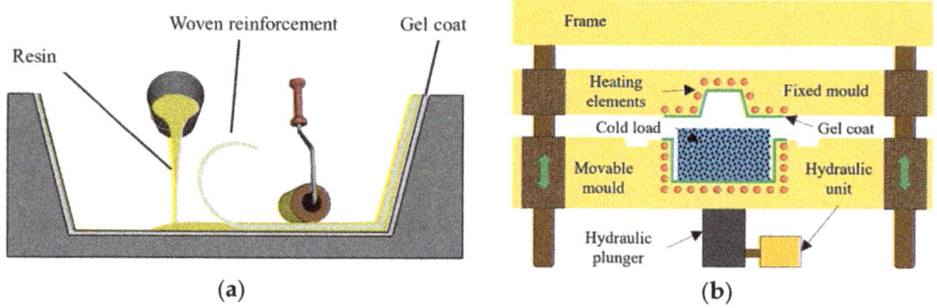

Figure 1. Schematic representation of the (**a**) HL and (**b**) CM process used for composite fabrication [1,19].

Injection molding (IM) is the commonly used method for fabricating polymer composites. The injection molding machine consists of a single screw extruder that melts the granular raw material and forces it into the mold cavity. A schematic representation of the IM technique employed for composite fabrication is represented in Figure 2. IM is a two-stage process. In the first step, the thermoplastic material is mixed with reinforcements in the form of particulates and chopped fibers (synthetic or natural) by compounding to obtain the pellets. In the second step, the pellets are injected into the mold cavity to obtain the components. A water jacket is provided around the mold to maintain the required temperature during the process. IM is a cyclic process, and very little scrap is generated during the process. The size and shape of the component and kind of raw material influence the IM cycle time [20].

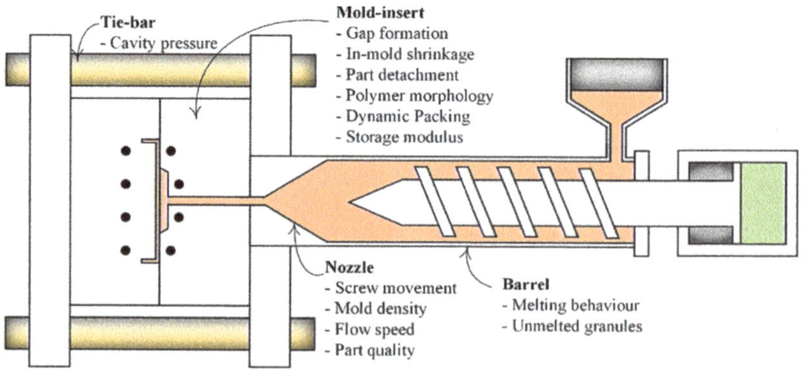

Figure 2. Schematic representation of the IM process used for composite fabrication [21].

Vacuum-assisted resin transfer molding (VARTM) is depicted in Figure 3. This process utilizes a vacuum bag and half mold to fabricate the composites. Reinforcements in the form of woven fabric or mats are placed on the bottom mold, which in turn is put inside the vacuum bag. The thermosetting resin is then drawn through the reinforcements placed in the vacuum bag to impregnate the fabrics, followed by curing using heating elements placed around the molds. Microwave heating sheets or thermal ovens can also be used to cure the fabricated composites. The process can be performed at higher pressures for fabricating higher-volume fraction composites that exhibit superior mechanical characteristics [22–24].

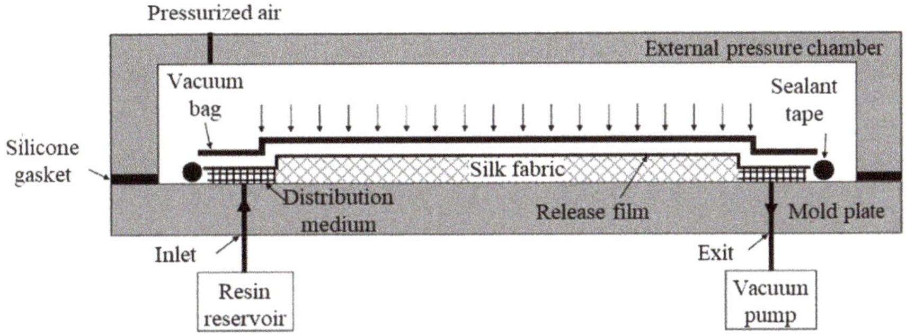

Figure 3. Schematic representation of the VARTM process used for composite fabrication [25].

RTM is a low-temperature and low-pressure process/technique employed for the fabrication of thermoset composites (Figure 4a). It is performed using two rigid closed molds and without the assistance of a vacuum. One of the molds is fixed, and the other can be moved using a hydraulic unit after laying the reinforcements. The curing agent and resin are then injected into the mold cavity under high pressure. Molds are surrounded by heating elements, which helps to cure the fabricated composites. RTM performed using higher pressure reduces the cycle time in the fabrication of intricate shape components [26,27].

FM involves meandering the impregnated fibers or tapes on a rotating mandrel along a prescribed path, such as helical, hoop, or polar (Figure 4b). Thermosetting resins are generally used to impregnate the fibers before winding. Fibers can be used in both wet and dry (pre-impregnated form) conditions. Wet winding involves unwinding the fibers from the roving and passing them through a resin mixture to impregnate them before they are wound on a mandrel with a certain orientation. Whereas the dry process makes use of fibers that have already been impregnated i.e., in semi-cured conditions, eliminating the onsite resin bath. Composites after winding are cured using an oven or autoclave or by using infrared radiation and removed from the mandrel. Higher fiber loading can be achieved using this process, which results in components with superior mechanical properties. This process can be easily automated and is suitable to produce only axisymmetric (cylindrical) components. Additionally, the processing behavior in filament winding is highly reliant on the properties of the resin [28,29].

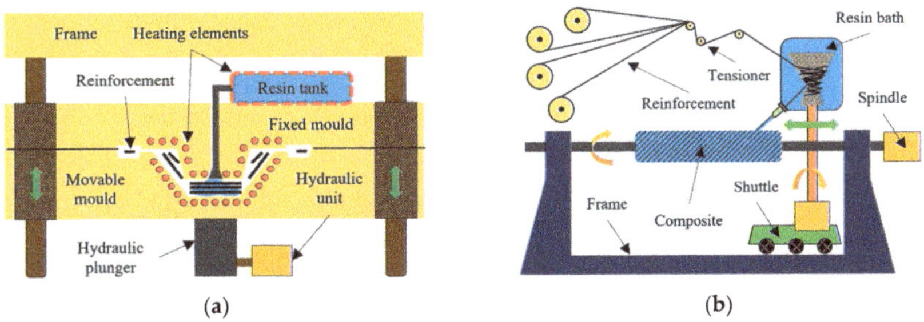

Figure 4. Schematic representation of the (**a**) resin transfer molding and (**b**) filament winding processes used for composite fabrication [1].

An autoclave molding process is a kind of open molding process, where molded parts are kept inside a plastic bag and cured utilizing vacuum, heat, and high-pressure inert gases inside an autoclave (a combination of an oven and pressure chamber). The autoclave molding process is shown in Figure 5. In the first step, prepregs of the desired component are firmly placed on the mold coated with a mold-releasing agent. Cores and inserts can also be used during the process. Later, the whole assembly is covered by a plastic vacuum bag before curing in an autoclave followed by vacuuming out the air resulting in denser and desired void-free components. High pressure is maintained throughout the curing process; however, the vacuum is maintained only during the early stages of the process. A high-temperature thermosetting resin, such as epoxy resin, can be processed using this process. The high cost and limited part size due to the use of autoclave machines and prolonged curing cycles are the limitations of the autoclave process [30–33].

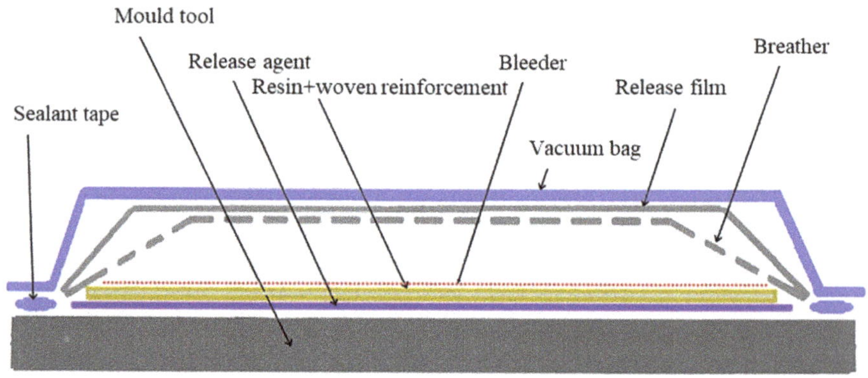

Figure 5. Schematic representation of autoclave molding process used for composite fabrication [19].

Pultrusion is the oldest technique to produce fiber-reinforced composites of a constant cross-section, employing closed heated dies. The pultrusion process is illustrated in Figure 6. The shape of the components produced relies on the die design used in the process. It is the simple, efficient, flexible, and most cost-effective technique for fabricating continuous-fiber structural composites with a constant cross-section. A creel for feeding the fiber, a resin reservoir, forming dies, a machined die with a temperature control feature, a puller, and a blade for cutting the product from the continuous system make up a pultrusion machine.

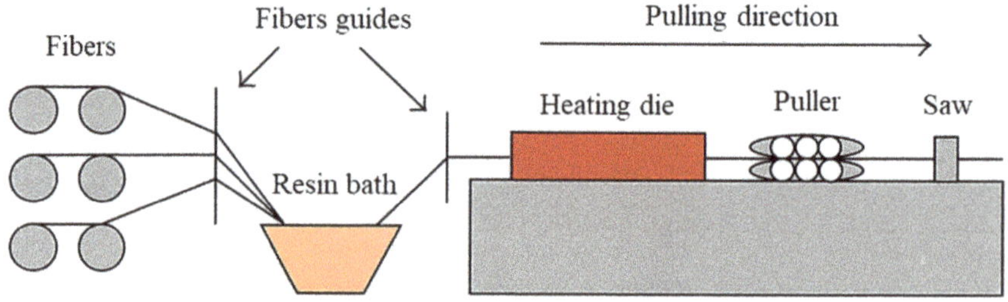

Figure 6. Schematic representation of pultrusion process used for composite fabrication [25].

Fiber forms (tape, woven, and/or mat) can be impregnated using both thermosetting and thermoplastic resins. Impregnated fibers are passed through preforming guides to remove excess resin and entrapped air, followed by passing through the heated die for curing. A sawing system is used to cut the cured composites according to the requirement. Resin-bath pultrusion and resin-injection pultrusion are the commonly used pultrusion techniques. High-fiber loading, as compared to the HL, process can be achieved in this process resulting in products with higher stiffness and strength [34–36].

AM belongs to the group of advanced manufacturing techniques where the components are fabricated in a layer-by-layer approach directly from computer-aided drawing without the use of tools or dies [37]. Various AM techniques can be used to fabricate the composite materials without using any molds and by selectively depositing the fibers [38]. Various AM techniques are illustrated in Figures 7 and 8.

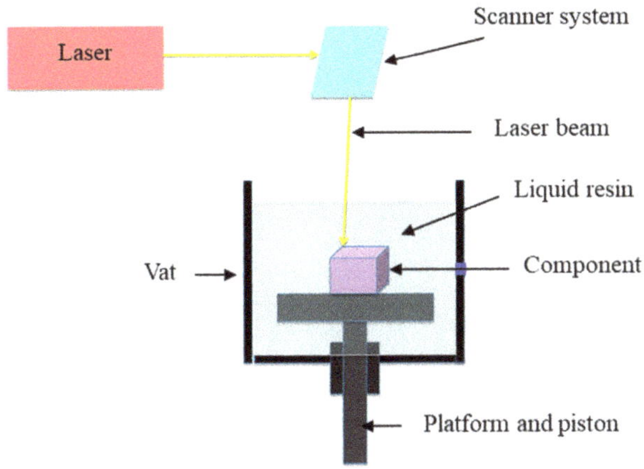

Figure 7. Schematic representation of stereolithography AM process used for composite fabrication [39].

A selective laser sintering (SLS) process is used to fabricate composites made of thermoplastic powder and short fibers using a low-power laser. Composites with complex geometries can be easily fabricated using the SLS process without using support structures [40]. The stereolithography (SLA) AM process utilizes a photosensitive resin along with a laser to fabricate the components. The photosensitive resin is mixed with particles, short fibers, continuous fibers, and woven fabric during fabrication. The volume fraction of the reinforcements is limited to 20 vol.% since a higher volume fraction of reinforcements may lead to non-uniform dispersion and inferior mechanical properties. Components after printing are cured using a UV light oven [41]. Fused deposition modeling (FDM) is an extrusion-type of AM method where the filament made of thermoplastic material is melted (up to softening temperature), extruded, and deposited using a heated nozzle (Figure 8). Multi-material deposition can also be achieved using this process utilizing multiple nozzles [41].

Continuous fiber fabrication and liquid deposition modeling (LDM) are the two common types of FDM processes. The process utilizes pre-impregnated continuous fibers with a thermoplastic matrix to fabricate the composites [43]. A matrix in liquid form is reinforced with discontinuous reinforcements and whiskers to fabricate the composites in the LDM type of the FDM process [41]. Other less commonly used AM processes to fabricate composites are three-dimensional printing and laminated object manufacturing [1]. Techniques adopted for the fabrication of composites reinforced by various artificial and natural fibers for engineering applications are discussed in detail in the following sections.

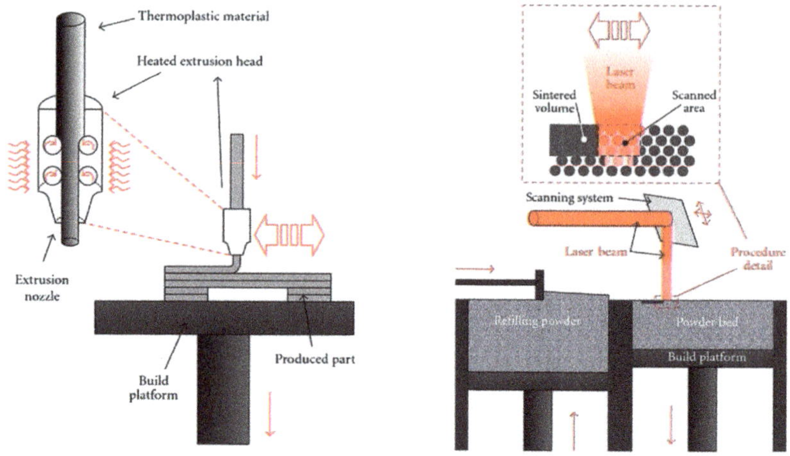

(a) **(b)**

Figure 8. Schematic representation of (**a**) FDM and (**b**) SLS AM processes used for composite fabrication [42].

3. Fabrication of Commonly Used Structural Composites

3.1. Aramid Fiber and Its Hybrid Composites

Aramid fibers (AFs) are slowly replacing glass fibers (GFs) to manufacture composites for engineering applications in the last decade because of their distinctive properties, such as low density, abundant availability, good thermal stability, high toughness, chemical solvent resistance, nonconductive property, impact and abrasion resistance, etc. These fibers exhibit excellent mechanical characteristics compared to steel and GFs on a weight-for-weight ratio. AFs are susceptible to ultraviolet radiation, absorb moisture, and exhibit low compression characteristics. AFs are extensively utilized in the production of insulated composites for military purposes, aircraft components, fire-resistant clothing, and armored vests [44].

AFs in the plain-woven fabric, along with natural kenaf fibers, were used to fabricate hybrid epoxy laminates using the HL technique. External pressure using dead weights (about 10 kg) was used to cure the composite at ambient conditions. The maximum fiber content of AFs in the study was limited to 53.64%. Fabricated composites were evaluated for their ballistic impact performance. Superior ballistic performance was observed in the composites with a lower volume fraction of kenaf fibers. Fabricated hybrid composites can be used for ballistic resistance applications, such as vehicle spall-liners [17]. Hybrid epoxy composites combined with aramid and semi-carbon fibers were also fabricated using the HL technique. Five different types of laminates were fabricated varying the aramid layer from zero to four. Aramid (Kevlar 49) and semi-carbon fibers were used in a twill form. Samples were cured overnight in ambient conditions. The reinforcement volume fraction was limited to 40%. Fabricated composites were evaluated for thermal and mechanical characteristics. Results indicate that with the increasing semi-carbon fiber content in the hybrid composite decreases the tensile strength and elastic and flexural modulus but decreases the rate of burning [45]. Aramid and natural (palm) fibers were successfully reinforced in an epoxy matrix to fabricate hybrid composites employing the HL technique. Reinforcement content was varied from 0 to 15 wt.% and was used in equal ratios (i.e., aramid:palm). Fabricated laminates were evaluated for various mechanical characteristics, namely tensile, flexural, impact, hardness, and water absorption tests. The flexural modulus of the fabricated composite decreases at the highest fiber content of 20 wt.% due to ineffective load transfer from one to another end of the composite. The water absorption rate was found to be highest for the composite reinforced with uncoated natural fibers due to voids [46]. Bio-degradable composites were successfully manufactured using bagasse/epoxy resin and AFs via an HL technique utilizing steel molds. Bagasse fibers were dried in sunlight, followed by chopping in a ball mill and washing using water to remove the pulps. Bagasse fibers were used in 40, 50, and 60 by wt.% of an epoxy matrix, while the aramid fiber weight percent was restricted to 5 wt.%. Laminates were dried at ambient temperature for 24 h and evaluated for dynamic mechanical properties, such as tensile, flexural, and impact strength. Superior mechanical properties were obtained in the composites reinforced with treated fibers [47].

The CM technique was employed to produce hybrid vinyl ester (VE) composites reinforced with coir fiber, coconut shell powder, and AFs (Kevlar 29). Coir fiber and coconut shell powder were used in as-received and surface (alkali and silane)-treated conditions. The AF content varied from 5 to 15 w.t%. The maximum fiber loading, including all types of reinforcement, was limited to 35 wt.%. Laminates were prepared by pressing the calculated wt.% of reinforcements into the mat using two chromium-coated mild steel molds, followed by resin pouring and rolling to remove entrapped air. Later, the mold was closed to cure the laminates using ambient temperature for a day under 1 MPa pressure. Prepared composites were studied for hardness, tensile, and flexural characteristics. Alkali and silane treated fiber-reinforced composites exhibited enhanced mechanical characteristics. The formation of silanol bonds and hydrogen bonds on fiber surfaces due to silane treatment enhances the adhesion between the filler–matrix, resulting in superior properties. Composites reinforced with treated fibers may be used in automobiles for structural applications [48]. Ultra-high-molecular-weight polyethylene (UHMWPE) is combined with the polymeric filler polytetrafluoroethylene (PTFE) and AFs using the CM technique. UHMWPE in powder form is mixed with reinforcements in the presence of ethanol to obtain a homogeneous mixture followed by heating in an oven at 79 °C to get rid of the ethanol. The molding temperature, pressure range, and time were set to 150 °C, 100–300 bar, and 5 min, respectively. Out of 5 min, heating was performed for 3 min followed by 2 min of cooling. The AF weight content was varied between 2 and 5 wt.%. Tribological tests were conducted to analyze the wear performance of fabricated composites. Composites manufactured at a higher molding pressure exhibited superior wear characteristics due to enhanced consolidation in the microstructures. Among the analyzed composites, the aramid fiber-reinforced composite demonstrated superior wear resistance [49]. Hygro-thermal degradation and mechanical studies were performed on E-glass woven roving and AF-reinforced hybrid epoxy composites fabricated via the CM technique. The AF content was varied from 9 to 13 wt.% [50]. Polyetherimide hybrid composites were manufactured utilizing woven CFs, GFs, and AFs (Kelvar 29) via a CM technique and evaluated for mechanical properties. Polyetherimide in a granular form was immersed in dichloromethane to fabricate different laminates with varying reinforcement contents. Composites were compression-molded at 400 °C and 8 MPa followed by cooling for 4–5 h maintaining the same pressure [51].

Short AFs were reinforced with polypropylene (PP) and thermoplastic elastomer ethylene-propylene-diene using the IM technique and evaluated for mechanical characteristics. AFs (Twaron 1488) used in the study had a diameter and length of 12 µm and 6 mm, respectively. The addition of AFs increases the impact strength of the PP matrix composites; however, a declining trend was observed for the ethylene-propylene-diene matrix composite. AFs display greater affinity towards a PP matrix, resulting in increased impact performance [52]. A polyamide (PA) matrix was reinforced with flakes and long-fiber granules obtained from aramid pulp to produce prepregs for the IM technique. First, compounding was performed employing a twin-screw extruder to produce the prepregs. The AF weight was varied from 5 to 20 wt.%. After compounding IM was performed to fabricate the composites. Initial materials during both stages were dried for 4 h at 80 °C using an oven. Fabricated samples were evaluated for tensile, flexural impact, and tribological characteristics. The authors successfully demonstrated a unique way of fabricating aramid fiber-reinforced PA prepregs for the IM technique. Fabricated composites showed decent tribological properties, which make them suitable for manufacturing components, such as gear wheels, deflection pulleys, etc., subjected to tribological loads [53]. Alumina nanoparticle-filled epoxy matrix and AF (Kevlar 49)-reinforced composite laminates were fabricated using RTM. Nanoparticles by 40 wt.% were dispersed in epoxy resin using a SpeedMixerTM at 2500 rpm. Laminates were fabricated using 14 layers of Afs, which was equivalent to 41% by volume of epoxy resin. RTM was performed using 3 bar injection pressure at 50 °C. Composites were first cured in the mold at 80 °C for 120 min, followed by secondary curing at 150 °C for 180 min in a forced convection oven for demolding. Fabricated composites were subjected to a particle burnout technique to measure the nanoparticle distribution. Nanoparticle distribution was found to be non-homogenous within the laminate, and a higher particle concentration was observed next to the entry of the epoxy matrix. Increasing the resin flow tie increases the concentration of nanoparticles in the fabricated laminate [54]. VARTM was employed to fabricate epoxy composites reinforced with carbon nanotubes (CNTs)/AFs (Kevlar-29). The diameters of AFs and CNT were 12 μm and 13 nm, respectively. First, purified CNTs were used to prepare polyethylene oxide nanocomposite films. Next, epoxy was heated to dissolve the nanocomposite film and disperse CNTs into the epoxy resin. Later, AF preforms were soaked in CNT-dispersed epoxy to fabricate the composites. Fabricated composites were evaluated for flexural characteristics. The authors successfully demonstrated a new fabrication technique to produce CNT/AF-reinforced epoxy composites. Incorporating CNTs enhances the flexural performance of hybrid composites [55].

Hybrid UHMWPE composites were fabricated by reinforcing AFs (Kevlar 29) in fabric form for vehicular ballistic protection using the autoclave molding technique. AFs were initially dried using an oven for 4 h at 105 °C before autoclaving to eliminate moisture. Ethylene vinyl acetate adhesive films were placed in between AFs and molded using the autoclave technique. The process was performed by varying the pressure from 6 to 24 bar and temperature from 80 to 110 °C for 55 min. Composites were subjected to a consolidation process at 24 bar and 110 °C. Later, the autoclave temperature was reduced to 30 °C along with the pressure. Fabricated composites were evaluated for dynamic mechanical properties and ballistic performance. Good adhesion was obtained between the AFs and matrix resin. The highest mechanical and ballistic performances were exhibited by the composite reinforced with 25 vol.% of AFs. Fabricated composites can be used for fabricating vehicle armors [56]. Graphene oxide-reinforced AF/epoxy composites were successfully fabricated using autoclave forming technology. The graphene oxide content was varied from 0.1 to 0.4 wt.%. After dispersing graphene oxide in the epoxy matrix, the mixture is vacuumed at 45 °C for 30 min before composite fabrication. A high-pressure spray gun was used to soak the aramid fiber before molding. Autoclave molding was performed at 150 °C and 0.4 MPa for 2 h. Fabricated composites were evaluated for their mechanical properties. The addition of graphene oxide significantly enhances tensile characteristics of AF-reinforced composites by inhibiting the stress concentrations introduced by the spaces between the fiber and matrix due to poor adhesion [57].

The FDM additive manufacturing technique was utilized to fabricate AF-reinforced nylon composites. A nylon filament reinforced with 2 wt.% short AFs was fabricated using a twin-screw extruder. The layer thickness, extruder temperature, infill part density, and raster angle varied in the range of 0.2–0.4 mm, 280–300 °C, 70–90%, and 0–90 degrees, respectively. A layer thickness of 0.4 mm, extruder temperature of 300 °C, infill part density of 90%, and raster angle of 90 degrees were identified as the optimum parameters for printing. Fabricated composites were assessed for tensile, flexural, compression, and impact performance. The layer thickness and raster angle significantly influence the mechanical performance of the printed samples. A uniform distribution of AFs throughout the nylon matrix was confirmed using scanning electron micrography [58]. A Multi Jet Fusion (MJF) AM technique was used to print high-strength AF/PA12 composites. Composites were printed using AF/PA12 composite powder and used in the as-received condition. PA12 powder was combined with short AFs utilizing a mechanical mixer. The AF content varies from 0 to 14 wt.%, and the mean fiber length was at the limit of 0.5–1.2 mm. The energy input, build platform temperature, layer thickness, and recoating speed were considered as process parameters for printing. Samples were printed using a layer thickness and build platform temperature of 20 µm and 220 °C, respectively. After printing, samples were naturally cooled to ambient temperature followed by cleaning using bead blasting [59].

3.2. Carbon Fiber and Its Hybrid Composites

CFRP composites find their applications in many engineering sectors, such as automobile, aviation, marine industries and civil engineering, wind-turbines, sports equipment, and robotics due to their unique properties, such as excellent specific mechanical properties, lightweight property, high damping ability, good dimensional stability, and resistance to the corrosive environment [60,61]. A carbon fiber can be reinforced with a variety of matrices, such as polymer, metal, ceramic, or carbon [62]. CFRP can be produced using various manufacturing techniques, such as vacuum bagging, CM, FW, HL, or AM techniques [63,64]. Even though CFRP finds many engineering applications, the fabrication costs are exceptionally high and the machining of CFRPs using conventional cutting tools is not economical [65,66].

The HL method was employed to fabricate CFRP using bisphenol-A epoxy matrix and CNTs. CFs were used in the fabric form and procured from Toray Industries Inc., Tokyo, Japan. Fibers were first treated with acetone to remove impurities, followed by treatment with nitric acid to enhance the wettability during processing. CFs were carefully cleaned in deionized water before being vacuum-dried for 12 h at 80 °C. Composites were produced via the HL method, followed by pressing at 150 °C and 7.4 MPa for 150 min. The CF content was limited to 60 wt.%. Fabricated composites were evaluated for their electrical and mechanical properties [67]. Multi-wall CNTs (MWCNTs) and CF-reinforced epoxy (bisphenol-A) nanocomposites were successfully fabricated using the FW technique. The MWCNT content was varied from 0 to 1 wt.% of the matrix weight. First, MWCNTs were dispersed in an epoxy matrix using a sonication process, followed by ball milling for 2 h at 250 rpm. The mixture was later used to fabricate unidirectional CF-reinforced composites using the FW technique. CF loading was limited to 60 wt.% of the epoxy resin. Composites were first dried at 120 °C for 120 min, followed by post-curing at 180 °C for 180 min. Fabricated composites were tested to evaluate their fracture toughness and mechanical properties. The crosslink density of the epoxy matrix was significantly enhanced by the presence of amino-functionalized-MWCNTs. This crosslinking between the filler and matrix enhances the interfacial strength of the composite along with the fracture toughness [68].

CM was employed to fabricate phenolic nanocomposites reinforced with CFs and MWCNTs. MWCNTs were first dispersed in phenolic resin by them stirring for 60 min at 500 rpm. CF and MWCNT loading was restricted to 50 wt.% and 2 wt.%. BMC paste was first subjected to CM at 1 bar pressure and 100 °C for 30 min. Later BMC paste was subjected to the second stage of CM at 10 bar pressure and 170 °C for 5 h. Composites were post-cured for 120 min at 200 °C. Fabricated composites were tested for thermal and mechanical properties. The flexural and thermal performance of the fabricated composite material was found to increase up to 1 wt.% of MWCNTs and decrease at 2 wt.% of MWCNTs. Micrography results revealed that failure of the pure composite was due to brittle fracture; however, ductile fracture was observed in MWCNT-reinforced composites. CNTs were uniformly dispersed in the phenolic resin without agglomeration [69]. Epoxy (Bisphenol-A) composite was successfully fabricated using CFs in a fabric form and MWCNTs using a CM technique. MWCNTs were ultrasonically dispersed in the epoxy resin before applying them on carbon fabric. The diameter and length of MWCNTs used in the study were in the range of 10–20 nm and 10–30 µm, respectively. The highest MWCNTs and carbon fiber loading were limited to 0.75 and 64 wt.%, respectively. MWCNTs were dispersed in acetone and mixed with epoxy, followed by sonication at 50°C for 60 min. Later, hardener was added and mixed thoroughly using a magnetic stirrer. Six layers of carbon fabric were stacked in a different orientation to fabricate laminate of a 1.5 mm layer thickness using the CM technique. Fabricated composites were cured using an autoclave under a vacuum and tested for various mechanical and thermal properties. The micro-crack bridging effect exhibited by nanotubes leads to the enhanced tensile and flexural performance of the composites. Uniform dispersion of MWCNTs was observed at a lower weight fraction, however, agglomeration was witnessed at a higher MWCNT content [70].

RTM was employed to successfully fabricate thick CFRP/epoxy composites. Composites were fabricated using 39 layers of CFs in different orientations. CF loading was limited to 55 wt.% of the matrix resin. The epoxy resin was injected into the molds at an injection speed of 2 l/m and 20 bar. Fabricated composites were tested for evaluating porosity and mechanical characteristics. The authors tried to propose optimum RTM process parameters to fabricate a 20 mm-thick CFRP plate. Performing compaction by means of compressed air enhances the aesthetic aspect and porosity of fabricated composites. Tested composites exhibited comparable mechanical characteristics to those fabricated via autoclave molding [71]. Hybrid epoxy composites reinforced with CFs and GFs were fabricated by employing the vacuum bagging technique. Composites were evaluated for various mechanical properties, such as hardness, tensile strength, and modulus. Bi-directional GFs and CFs in the woven mat form of 200 GSM were used for composite fabrication. The fiber weight percentage was varied from 15 to 60 wt.%. First, the laminates were prepared using the HL technique at ambient temperature, followed by vacuum bagging at 0.1 bar to remove the trapped air. Composites were post-cured in three stages, namely at 50 °C for 30 min, at 65 °C for 45 min, and at 75 °C for 60 min. Enhanced mechanical performance was obtained at a higher wt.% of fiber reinforcements. CF-reinforced composites exhibit enhanced ductile characteristics compared to the composites reinforced with GFs [72].

Hybrid PP composites mixed with short carbon fibers (SCFs) and short glass fibers (SGFs) were successfully fabricated by employing extrusion compounding and IM techniques. Fiber loading in hybrid composite fabrication was restricted to 25 wt.%. A twin-screw extruder was employed to manufacture PP composites using GFs and CFs in the roving form. The extrudates were cooled and pelletized, followed by IM, using a twin-screw injection molding machine at 230 °C [73]. PP composites reinforced with short CFs were fabricated using an extrusion and IM process. CF loading was limited to 40 wt.%. The CF diameter and mean length were in the ranges of 7–9 μm and 150 μm, respectively. Six different types of composites were manufactured by varying the reinforcing CF content. Initially, mechanical mixing is performed to mix the constituents, followed by melt-mixing using a twin-screw extruder at 210 °C, followed by pelletization. The pellets were injection-molded at 230 °C to fabricate the composites. The composites were subjected to mechanical testing to determine the tensile strength and modulus. CFs were randomly oriented with respect to the injection direction. More fiber breakage and agglomeration were noticed at higher SCF loading. The composite exhibits inferior ductility characteristics due to the presence of hollow spaces at fiber ends and lower interfacial strength among the constituents [74].

SCF-reinforced polycarbonate composite material was successfully printed employing the fused filament fabrication (FFF) technique. Initially, filaments of 2.85 mm were fabricated using a single screw extruder at 270 °C and a 25 mm/s speed. Constituent materials were dried in a furnace for 120 min at 90 °C to eliminate the water content before extruding them. The SCF content was varied from 3 to 10 wt.%. Samples were printed in a ULTIMAKER3 printer by maintaining a bed temperature of 107 °C, extrusion temperature of 270 °C, infill density of 100%, and layer thickness of 0.2 mm, with varying printing speeds in the range of 25–75 mm/s. Printed samples were subjected to mechanical testing to determine tensile, flexural, compression, and micro-hardness properties [75]. Continuous fiber-reinforced nylon composites were printed employing an FDM-based Markforged Mark One 3D printer. Carbon, glass, and Kevlar fibers were used as reinforcements for printing hybrid composites. A nylon filament of 1.75 mm was stored in a closed box to avoid moisture absorption before printing. The diameter of carbon, Kevlar, and glass fibers was 8 μm, 12 μm, and 10 μm, respectively. Samples were printed in two stages. In the first stage, nylon was printed on a built platform kept at ambient temperature. In the second stage, fibers were oriented as per the desired direction and printed on a nylon layer. Printed composites of glass and Kevlar consisted of 32 layers, while carbon fiber-reinforced composites consisted of 26 layers. Samples were printed using fibers in concentric and isotropic patterns. Both the matrix and reinforcements were extruded at 263 °C using distinct print heads. The printed composites were analyzed for both tension and flexure performances [76]. An SLS additive manufacturing technique was successfully employed to print CFRP composites using PA12 in powder form. CF loading was varied from 30 to 50 wt.%, and the surface of the fibers was modified using an oxidation modification technique to enhance adhesion with the matrix material. The composite powder was prepared using a suitable technique, and the samples were printed using a continuous wave CO_2 laser. The following process parameters were employed during printing: laser beam speed—500 mm/s; laser power—22 W; layer thickness of powder—0.1 mm. Manufactured composites were examined for evaluating their mechanical and thermal characteristics [77]. An SLS additive manufacturing technique was employed to print recycled PA12 composites reinforced with milled CFs up to 30 wt.%. Constituent materials were first dried using an oven for 24 h at 60 °C, followed by manual mixing using a plastic bag. The mixture was then compounded utilizing a twin-screw extruder maintaining a constant temperature setting and screw speed. Extruded material was pelletized and dried again for a day at 60 °C. Filaments were fabricated using a single screw extruder, followed by printing using Ultimaker 2. Samples were printed at a printing speed of 60 mm/s, build platform temperature of 240 °C, and layer thickness of 0.2 mm, with 100% fill density. Printed samples were evaluated for mechanical and thermal characteristics [78].

3.3. Glass Fiber and Its Hybrid Composites

Glass fibers are widely used in the fabrication of structural composite materials because of their high strength, stiffness, and durability. GFs can be used in a variety of forms, such as short fibers, continuous fibers, fabric, or mat forms to fabricate the composites. GFs can be reinforced with many polymers, such as epoxy, phenolic, polyester, vinyl ester, etc. [62]. GF-reinforced composites have several advantages over traditional materials, such as metal and wood. They are lightweight and corrosion-resistant and have high strength-to-weight ratios. They are also highly customizable, as the properties of the composite can be tailored to meet specific application requirements. These materials are used in many applications, namely aerospace, automotive, marine, and construction industries. They are commonly used to make parts, such as body panels, structural components, and insulation [79–81].

GFRP composites were fabricated using both HL and VARTM followed by a comparative analysis of mechanical and thermo-mechanical performance for wind turbine blade applications. Bi-directional E-glass woven fibers were utilized as reinforcements with an epoxy matrix. Specimens with 10 layers of GFs were fabricated using HL followed by curing under a 25 kg weight for 24 h. Similarly, specimens of 10 layers were also fabricated using VARTM at room temperature for a comparative analysis [82]. GFRP hybrid composites were fabricated using an HL and vacuum infusion (VI) technique for marine vessel applications. The outer layer of hybrid composites was fabricated using the HL technique and the interior layers using the VI technique at a different orientation. Epoxy vinyl ester and plain vinyl ester were used as matrix material for VI and HL techniques, respectively. Fabricated samples were evaluated for tensile, compression, and in-plane shear characteristics [83].

Short glass fibers and calcium carbonate-reinforced polyester composites were fabricated from BMC using a CM technique. Calcium carbonate at 55 wt.% was used to first fabricate BMC. Fibers of two different lengths were used, and the mean length of chopped glass fibers was 0.4 mm and 6.4 mm. Composites were fabricated by employing the following process parameters: mold temperature of 149 °C; mold pressure of 5.5–6.9 MPa; cure time of 2 min. Fabricated samples were evaluated for impact characteristics using the Charpy test for aerospace applications [84]. Hybrid epoxy composites of glass and jute fibers in woven form were successfully fabricated using the CM technique. The E-glass and jute reinforcement thicknesses were 0.7 and 1 mm, respectively. After CM, first, a curing pressure of 15 kg/cm^2 was applied for 10–15 min to enhance wettability, followed by post-curing at room temperature for 24 h. Hybrid composites were evaluated for mechanical characteristics in terms of flexural and impact strength followed by numerical analysis to evaluate the predictive flexural response [85].

Glass fiber-reinforced polyester pipes were manufactured using the FW technique and evaluated for tensile characteristics. The inner layer of the composite was fabricated using a glass fiber mat of 450 GSM using polyester resin on a cylindrical mandrel. After curing the first layer, polyester resin-impregnated continuous E-glass woven fabrics were wounded to fabricate the structural layer over the inner layer. The highest fiber and matrix weight fraction was limited to 79.5% and 34.4%, respectively [86]. Hybrid polyester composites reinforced with woven jute/glass fabric were fabricated using the HL technique. The weights of woven glass and jute fabrics were 360 and 320 GSM, respectively. Laminates were first cured using moderate pressure for 60 min, followed by secondary curing without any pressure for 48 h at room temperature. Six types of laminates were fabricated using 10 layers of reinforcements, and the highest fiber loading was limited to 42 wt.%. Fabricated composites were tested for tensile, flexural, and interlaminar shear characteristics [87].

Vacuum Assisted Resin Injection (VARI) was employed to fabricate an unsaturated polyester composite reinforced with E-glass fabrics. Laminates were manufactured using three layers of glass fabrics, followed by 40 h of curing under ambient conditions. Fabricate composites were examined to evaluate the influence of water immersion on mechanical performance [88]. Glass fiber-reinforced unsaturated polyester resin composites were manufactured using the RTM process. Before the RTM process, the fibers were treated with a coupling agent for 10 min and dried under ambient conditions for 48 h, followed by oven drying under a vacuum for 24 h. Laminates were fabricated using 12 layers and mold was wrapped with a rubber dam to avoid resin leakage. Fibers were impregnated at 40 °C under 46 kPa pressure using nitrogen gas. Later, the laminates were cured by increasing the mold temperature to 85 °C for 60 min. Fabricated composites were tested for void contents and flexural characteristics [89].

An SLA process was employed to print short and continuous fibers composites reinforced with photo-curable epoxy resin. Samples were printed utilizing a Nobel 1.0 SLA 3D printer manufactured by XYZ Printing, Inc. using a laser wavelength of 405 nm. Reinforcements in the form of glass powder, a chopped glass strand mat, and fiberglass fabric were coated with silane before printing for enhancing the interfacial bonding. Glass powder-reinforced composites were fabricated by varying the reinforcement content up to 55 wt.% with a layer thickness of 0.1 mm. Specimens were cured for 10 min using an exposure unit Good quality specimens were obtained up to 50 wt.% of glass powders. Intricate shapes of good quality were produced using glass powders up to 10 wt.%. A chopped strand mat was used to fabricate short glass fiber-reinforced composites employing the same printing parameters. The reinforcement content was maintained at 1 wt.% of the epoxy resin. Continuous glass fiber specimens were fabricated with a layer thickness of 0.1 mm by submerging the fiberglass fabric in a vat containing a photocurable resin. Specimens were successfully fabricated with a few minute voids of less than 1 mm in size using glass powder and continuous fibers. However, the fabrication of short fiberglass composites was not successful due to non-consistent reinforcement dispersion and the existence of large holes [90].

Three-phase GF-reinforced composites were fabricated using the SLS additive manufacturing technique, followed by infiltration using epoxy, for insulating purposes. GF was surface-treated using silane (KH-550) before reinforcing it with a phenol formaldehyde resin. Phenol formaldehyde was dissolved in ethanol and mixed with GF to prepare powder for the SLS process using the ball milling process. Ball milling was performed at 300 rpm for 60 min and constituents were dried for 24 h at 50 °C, followed by crushing and sieving, to obtain the suitable powder material. GF loading was varied from 60 to 80 by vol.% of the matrix resin. Samples were printed using a continuous-wave CO_2 laser in an HK S320 SLS system. Samples were printed with the following optimum conditions: build plate temperature of 65 °C; laser power of 14 W; scanning speed of 3500 mm/s; layer thickness of 0.1 mm. Finally, the fabricated samples were infiltrated with epoxy to fill the pores, followed by curing in an oven. Fabricated composites were evaluated for mechanical and electrical properties [91]. An FDM additive manufacturing technique was employed to fabricate an SGF-reinforced acrylonitrile butadiene styrene polymer composite. SGF loading was limited to 30 wt.%. ABS pellets were first dried in an oven for 120 min at 120 °C to eliminate moisture. After drying, the ABS pellets were mixed with the required quantity of SGFs and compounded using a twin-screw extruder at 50 rpm and 225 °C. Compounded pellets were used to fabricate filaments of 1.75 mm with a twin-screw extruder using a temperature range of 200–230 °C. Samples were printed adopting the following printer settings: infill density of 100%; layer thickness of 0.1 mm; build plate temperature of 80 °C, and extruder temperature of 240 °C. Fabricated samples were tested for mechanical performance evaluations [92].

3.4. Natural Fiber and its Hybrid Composites

A natural fiber-reinforced polymer composite is a material in which a polymer matrix is reinforced with natural fibers. Natural fibers, such as jute, hemp, flax, sisal, and kenaf, are commonly used for reinforcing the polymer matrix [93]. Natural fibers enhance mechanical properties, such as strength, stiffness, and toughness. Natural fibers have been extensively used in polymer composites in recent years because they are renewable, low-cost, lightweight, and environmentally friendly [94]. Various techniques can be adopted to fabricate these types of composites. NFRP composites have many advantages over traditional composites made with synthetic fibers. They have a lower environmental impact, are biodegradable, and have good mechanical properties. Additionally, the use of natural fibers can reduce the weight of the composite, making it lighter and more fuel-efficient in certain applications. Natural fibers were generally surface-treated to enhance the adhesion with the matrix material before composite fabrication. The resulting composite material can be used for many purposes, such as automotive parts, construction materials, furniture, and packaging. Overall, natural fiber-reinforced polymer composites are a promising material that can offer a sustainable solution for various industries [95,96].

Hybrid epoxy composites reinforced with unidirectional sisal/banana fibers were fabricated using an HL Process. Reinforcements were surface-treated with a sodium hydroxide solution before composite fabrication. The reinforcement content was restricted to 30 wt.% of the epoxy resin. After fabrication, laminates were cured under a light load for 24 h at room temperature. Fabricated composites were drilled and evaluated for delamination characteristics [97]. An HL technique was employed to fabricate hybrid polyester composites using natural flax and synthetic GFs for wind turbine blade applications. Three different types of laminates were fabricated by varying the reinforcement contents. Fabricated composites were examined to estimate the mechanical characteristics and water absorption characteristics [98]. VE composites reinforced with surface-treated coir fibers were fabricated using the HL technique. Coir fibers were derived from the coconut husk procured from a local market. Samples were cured under a constant pressure of 5 MPa for 24 h at room temperature. The mean length and diameter of the fibers were 8 mm and 200 µm, respectively. Fabricated composites were evaluated for tensile and flexural characteristics [99].

The comparative mechanical and thermal characterization was performed on flax fiber/epoxy composites fabricated using HL and CM techniques. Reinforcements were used in mat form without any surface treatments. The reinforcement contents were varied from 24–30 by vol.% of the epoxy resin. HL was performed at room temperature, followed by two stages of curing. Samples were initially cured at 80 °C for 4 h and later at 120 °C for 2 h. CM was performed by varying the compression pressure in the range of 2–10 MPa and temperature from 80 to 120 °C. Superior characteristics were obtained for the compression -molded composites with 27 vol.% of flax fibers at 4 MPa pressure and 100 °C temperature [100]. Polylactic acid (PLA)-based bio-composites were fabricated using the CM technique using bamboo, cotton, and flax fibers in non-woven mat form.

The mean lengths of bamboo, cotton, and flax fibers were 126, 35, and 80 mm respectively. Constituent materials were first dried at 60 °C overnight before compression using a hot press for 15 min at 185 °C under 1250 psi pressure. Laminates were cured at 120 °C for 30 min using an oven. Fabricated composites were evaluated for mechanical (flexural and impact), thermal, damping, and acoustic characteristics and were to be used in automobile and construction industries. A PLA composite reinforced with cotton and bamboo fibers exhibited better performance than the commercially available material [101].

RTM was employed to successfully fabricate epoxy composites reinforced with unidirectional high-quality flax fibers in mat form. Composites were fabricated using three different volume fractions of the reinforcements. The highest fiber loading was limited to 48 by vol.% of the epoxy resin. Injection pressures of 2 bar and 1 bar were employed to fabricate the composite reinforced with high- and low-volume fractions of reinforcements, respectively. Resin was injected into the closed mold at 50 °C, followed by curing at 80 °C for 8 h. The prepared composites were tested to evaluate their mechanical properties [102]. Green PLA composites reinforced with bamboo and pineapple leaf fibers were fabricated using the IM technique. Natural fibers were surface treated using three different solutions to enhance the wettability before fabrication. Fiber loading was limited to 10 wt.%, and the mean length of the fibers was 4 mm. First, PLA pellets were dried in an oven at 70 °C for 6 h to remove the water content. Later, IM was performed by considering the following process parameters: barrel temperature range, injection pressure, and speed of 125–165 °C, 90 bars, and 60 mm/s, respectively, followed by a 25 s cooling cycle. Fabricated composites were evaluated for various mechanical properties, such as tensile, flexural, compression, and shear properties [103].

Bamboo fiber-reinforced PP/PLA composites were successfully fabricated using the FDM additive manufacturing technique. Natural fibers were subjected to various surface treatments to enhance their wettability before printing. First, PLA composites were fabricated using IM by adopting the following conditions: IM temperature range of 180–185 °C; injection pressure of 60 MPa; and holding time of 10 s. Next, PP/PLA/bamboo fiber composite filaments were manufactured using a single-screw extruder. Filaments were made of 20 wt.% bamboo fiber, 52.5 wt.% PP, and 22.5 wt.% PLA. Composites were printed using an FDM-based printer using an extrusion temperature of 180–200 °C, a printing speed of 40–60 mm/s, and a build plate temperature of 40–60 °C. Printed samples were evaluated for mechanical and thermal characteristics [104]. A hybrid AM technique (FDM in combination with shape deposition modeling) was employed to print natural-fiber granulated composites made of sugarcane, jute, ramie, banana, pineapple fiber, and seashell powder. The volume fraction of the reinforcements was fixed at 80 vol%. Fibers were first washed using distilled water, next treated with sodium hydroxide solution, and later dried at 80 °C for 180 min to eliminate moisture before milling using a ball miller to the size of 70 to 100 µm. The reinforcements were mixed in the required proportion and converted into a paste using epoxy resin. After the paste was obtained, the composites were printed using a specialized printing head using an FDM printer. Samples were printed by varying the number of layers from 6 to 12; however, the thickness of the composites was limited to 13 mm. Granulated composites were tested for evaluating their various mechanical properties [105]. An SLS additive manufacturing technique was successfully employed to fabricate bamboo flour/co-polyester composites. Bamboo flour contents of 20, 25, and 30 by wt.% were used to fabricate the composites. Bamboo floor and co-polyester powder were mixed in different ratios using a high-speed mixer for 10 min at 700–800 rpm and 50 °C to prepare the composite power for SLS. SLS was performed using a CO_2 laser with a 10.6 mm wavelength and a scan speed of 2000 mm/s. Printed samples were evaluated for mechanical properties in terms of tensile, flexural, and impact strength [106].

4. Mechanical Characterization of Fibrous Composites

Mechanical characterization of fibrous composites involves studying the behavior of composite materials under mechanical loading, which can provide insights into the strength, stiffness, and durability of the material. The mechanical properties of the fibers, the matrix material, and the interface between them are important in determining the overall mechanical behavior of the composite. The results of these tests can be used to design and optimize composite materials for specific applications, such as aerospace, automotive, and construction industries. The mechanical characterization of synthetic and natural fibrous composites is discussed in the following section.

4.1. Synthetic Fiber-Reinforced Polymer Composites

Jaiswal et al. [107] examined the effect of AF content and its orientation on the mechanical performance of the composites, such as tensile strength, modulus, and elongation at break. The results showed that increasing the AF content improved the mechanical properties of the composites, with a maximum tensile strength achieved at a 50:50 AF-to-polypropylene fiber ratio. Furthermore, the study found that increasing the fiber orientation angle resulted in decreased tensile strength and elongation at break but an increased modulus. Kandekar and Talikoti [108] investigated the effect of the number and position of AF strips on the torsional strength of the reinforced concrete beams. The experimental results indicated that the use of AF strips can significantly improve the torsional strength and stiffness of the reinforced concrete beams. Furthermore, the study found that placing the AF strips at the critical sections of the beam can improve the torsional behavior of the reinforced concrete beams more effectively than placing them at other locations. The addition of treated coir fiber/coconut shell powder and AFs improved the mechanical properties of the VE composites, with the highest mechanical properties achieved at a 10 wt.% fiber content. The study found that the addition of AFs had a more significant effect on the tensile and impact strength of the composites, while the addition of treated coir fiber/coconut shell powder had a more significant effect on the flexural strength [48]. Yahaya et al. [17] investigated the ballistic impact properties of woven kenaf-aramid fiber hybrid composites. The results showed that the addition of AFs improved the ballistic impact properties of the composites, including increased energy absorption and reduced back-face deformation. The composites with a higher percentage of AFs had better ballistic impact properties but at the cost of decreased tensile strength.

In a study by Badakhsh et al. [67], the effect of grafting CNTs onto carbon fibers (CFs) in CFRP composites was investigated. The study found that the grafting process improved the surface energetics of CFs, leading to better bonding between the fibers and the polymer matrix. As a result, the mechanical properties of the CFRP composites were enhanced. The study also suggested that optimizing the grafting process can achieve a balance among surface energetics, electrical conductivity, and mechanical properties. Tariq et al. [70] investigated the mechanical properties of multi-scale CNT/CF/epoxy composites. The incorporation of CNT into carbon fiber-epoxy composites enhances their mechanical properties by providing additional reinforcement at the nanoscale. The resulting composites exhibit improved stiffness, strength, toughness, and fatigue resistance. Gupta et al. [75] conducted a study to examine the impact of varying the CF content on the mechanical properties of a composite material. The results showed that the addition of CFs to the polycarbonate matrix improved the mechanical properties of the composite. The inclusion of 10 wt.% CFs increased the tensile strength and modulus by 52% and 82%, respectively, when compared to pure polycarbonate. Additionally, the incorporation of CFs also improved the flexural strength and modulus of the composite material. According to microstructure analysis, the CFs were uniformly dispersed throughout the polycarbonate matrix, which contributed to the enhanced mechanical properties.

Huang and Sun [88] investigated the effect of water absorption on the mechanical properties of glass/polyester composites. The results showed that water absorption significantly affected the mechanical properties of the glass/polyester composites. Specifically, the tensile strength, flexural strength, and impact strength of the composites decreased after immersion in water for various durations. The authors attributed this decrease in mechanical properties to the degradation of the resin matrix due to water absorption. The study also revealed that the extent of degradation was dependent on the duration of water immersion and the type of fiber reinforcement. Ahmed and Vijayarangan [87] studied the mechanical properties of jute and jute-glass fabric-reinforced polyester composites. The results of the study showed that the addition of GFs to the jute fabric improved the mechanical properties of the composites. Specifically, the jute-glass fabric-reinforced polyester composites had higher tensile strength, flexural strength, and interlaminar shear strength compared to the jute-reinforced polyester composites. The microstructure analysis showed that the water absorption caused swelling of the resin matrix, which led to the formation of voids and cracks at the interface between the fiber and the matrix. This resulted in a decrease in the interfacial adhesion and, hence, the mechanical properties of the composites. The influence of fiber length on the mechanical properties of GF-reinforced PA12 composites manufactured using multi-jet fusion printing was investigated by Liu et al. [109]. It was found that using longer fibers resulted in increased porosity in the composite parts. The addition of GFs with an average length of 226 μm significantly improved the ultimate tensile strength and tensile modulus of the composites in the direction of powder bed spreading. Specifically, the improvements were 51% and 326%, respectively, compared to pure PA12 specimens. Li et al. [91] evaluated the mechanical properties of electrical insulating composites made from glass fiber, phenol-formaldehyde resin, and epoxy resin using SLS technology. The results indicated that the bending and tensile strength of the composites increase by 30% and 42.8%, respectively, after being infiltrated with epoxy resin. Additionally, increasing the glass fiber content improves the flexural and tensile strength of the composite. These improved properties make the hybrid composites suitable for use in complex structural electrical insulation devices fabricated using SLS additive manufacturing technology, thereby broadening the materials and applications of this technology.

4.2. Natural Fiber-Reinforced Polymer Composites

According to research of Thomason [110], the transverse and shear modulus of jute fiber is significantly lower than its longitudinal modulus, leading to lower mechanical performance of NFRP composites compared to GFRP composites, mainly due to weaker mechanical properties in the transverse direction. This weakness can be attributed to the detachment of the outer layer and individual fibers in contact with the matrix during compounding and extrusion, which is not caused by a lack of physical and chemical compatibility between the fiber and matrix. A study by Ouali et al. [111] showed that a high-density polyethylene composite made with a 40% kenaf fiber mat had similar mechanical properties to a composite made with 40% discontinuous glass fiber and HDPE. Both composites had specific tensile strengths and moduli, as well as flexural strengths and moduli. Additionally, another process that involved coating flax fabric with adhesive, sandwiching it with polymer films, and compression molding resulted in a flax/PE composite with remarkable tensile strength and modulus values. According to a study by Couture et al. [112], the mechanical properties of flax/PLA and flax-paper/PLA composites were compared. Results showed that both composites had similar specific tensile properties, 252 MPa cm^3/g and 217 MPa cm^3/g, respectively, compared to composites made with woven glass fabrics and epoxy. The flax-paper composite also had an exceptionally high impact strength of 600 J/m compared to unreinforced resin of 15 J/m.

A summary of the various investigations performed on fibrous composites fabricated using different manufacturing techniques is presented in Table 1.

Table 1. Summary of the investigations performed and manufacturing methods adopted for the fabrication of fibrous composites.

Manufacturing Technique	Type of Matrix	Type of Reinforcement	Investigations Performed	Ref.
Autoclave molding technique	Epoxy	AFs	Mechanical properties	[57]
	UHMWPE	AFs	Dynamic mechanical properties	[56]
CM	Epoxy	GFs and AFs	Hygro-thermal degradation and mechanical properties	[50]
	Epoxy	MWCNTs and CF	Thermal and mechanical properties	[71]
	Epoxy	GFs and Jute Fibers	Flexural and impact properties	[85]
	Phenolic resin	MWCNTs and CFs	Thermal and mechanical properties	[70]
	Polyester	SGFs and Calcium Carbonate	Impact characteristics	[84]
	Polylactic acid	Bamboo, Cotton, and Flax Fibers	Mechanical, thermal, damping, and acoustic characteristics	[101]
	Polyetherimide	CFs, GFs, and AFs	Mechanical characteristics	[51]
	UHMWPE	PTFE and AFs	Wear performance	[49]
FDM	ABS	SGFs	Mechanical properties	[92]
	Nylon	AFs	Tensile, flexural, compression, and impact properties	[58]
	Nylon	CFs, GFs, and Kevlar Fibers	Tension and flexure performances	[77]
	PP/PLA	Bamboo Fibers	Mechanical and thermal characteristics	[104]
FDM and shape deposition modeling	Epoxy	Sugarcane, Jute, Ramie, Banana, Pineapple Fibers, and Seashell Powder	Mechanical properties	[105]
Fused filament fabrication	Polycarbonate	SCFs	Tensile, flexural, compression, and micro-hardness properties	[76]
FW	Epoxy	MWCNTs and CFs	Mechanical properties	[69]
	Polyester	GFs	Tensile characteristics	[86]
HL	Epoxy	AFs and Kenaf Fibers	Ballistic impact performance	[17]
	Epoxy	AFs and CFs	Thermal and mechanical characteristics	[45]
	Epoxy	AFs and Palm Fibers	Tensile, flexural, impact, hardness, and water absorption tests	[46]
	Epoxy	AFs and Bagasse Fibers	Tensile, flexural, and impact strength	[47]
	Epoxy	CFs	Electrical and mechanical properties	[68]
	Epoxy	Sisal and Banana Fibers	Delamination characteristics	[97]
	Hybrid VE	Coir Fibers, AFs, and Coconut Shell Powder	Hardness, tensile, and flexural characteristics	[48]
	Polyester	GFs and Jute Fibers	Tensile, flexural, and interlaminar shear characteristics	[87]
	Polyester	Flax Fibers and GFs	Mechanical and water absorption characteristics	[98]
	VE	Coir Fibers	Tensile and flexural characteristics	[99]

Table 1. *Cont.*

Manufacturing Technique	Type of Matrix	Type of Reinforcement	Investigations Performed	Ref.
HL and vacuum infusion	Epoxy VE and plain VE	GFs	Tensile, compression, and in-plane shear characteristics	[83]
Hand layup and compression molding	Epoxy	Flax Fibers	Mechanical and thermal characterization	[100]
HL and VARTM	Epoxy	GFs	Mechanical and thermo-mechanical properties	[82]
IM	PLA	Bamboo Fibers	Tensile, flexural, compression, and shear properties	[103]
	PA	AFs	Tensile, flexural, impact, and tribological characteristics	[53]
	PP	SGFs and CFs	Mechanical characteristics	[74]
	PP	SCFs	Tensile strength and modulus	[75]
	PP and ethylene-propylene-diene	Short AFs	Mechanical characteristics	[52]
MJF	PA12	AFs	Mechanical properties	[59]
RTM	Epoxy	AFs and Alumina Nanoparticle	Nanoparticle distribution	[54]
	Epoxy	CNTs and AFs	Flexural characteristics	[55]
	Epoxy	CFs	Porosity and mechanical characteristics	[72]
	Epoxy	Flax Fibers	Mechanical properties	[102]
	Polyester	GFs	Void contents and flexural characteristics	[89]
SLA	Epoxy	Glass Powder, Chopped Glass Strand Mat, and Fiberglass Fabric	Tensile characteristics	[90]
SLS	Polyester	Bamboo flour	Tensile, flexural, and impact strength	[106]
	PA12	Milled CFs	Mechanical and thermal properties	[78]
	Phenol formaldehyde	GFs	Mechanical and electrical properties	[91]
Vacuum-assisted resin injection	Polyester	GFs	Water immersion on the mechanical performance	[88]
Vacuum bagging	Epoxy	CFs and GFs	Hardness, tensile strength, and modulus	[73]

5. Conclusions

This review has provided an overview of the fabrication and mechanical characterization of fibrous composites for engineering applications. Fibrous composites offer a range of desirable properties, such as high strength, stiffness, and low weight, making them attractive for use in various engineering applications. The review highlights the importance of proper fabrication methods and their impact on the mechanical properties of the composites. The following are the critical observations made during the study:

- FRPCs were successfully fabricated using both conventional and additive manufacturing techniques. Conventional techniques, such as HL, CM, RTM, VARTM, IM, VI, VARI, FM, and autoclave forming, and AM techniques, such as FDM, MJF, SLS, and SLA, were used to fabricate FRPCs.
- AFs were reinforced with many polymers, such as epoxy, vinyl ester, UHMWPE, polyetherimide, polyamide, and polypropylene, during composite fabrication.
- Polymers, such as epoxy, phenolic resin, polypropylene, polycarbonate, polyamide, and nylon, were used to fabricate CFRP composites.
- Reinforcements were used in the form of short fibers, continuous fibers, woven fabrics, and powders for composite fabrication.
- AF, CF, GF, and natural fiber content varied in the ranges of 5–54 wt.%, 10–64 wt.%, 30–80 wt.%, and 20–80 vol%, respectively, for composite fabrication.

- The highest fiber loadings of 64 wt.% in CFRP composites, 80 wt.% in GFRP composites, and 80 vol% in NFRP composites were achieved by employing the autoclave forming process, SLA, and hybrid AM technique (FDM in combination with shape deposition modeling), respectively.
- HL and FDM were extensively used conventional and AM techniques, respectively, for fabricating AFRP composites.
- AM techniques, such as FFF, FDM, and SLS, were used for CFRP fabrication; however, reports regarding the utilization of other AM techniques were not found during the study.
- Glass fibers were also employed to fabricate hybrid composites by reinforcing them with various natural fibers and polymer resins, such as epoxy, polyester, VE, and phenol formaldehyde resin.
- NFRP composites were fabricated by reinforcing sisal, banana, jute, flax, bamboo, cotton, coir, pineapple leaf, etc., with epoxy, polyester, VE, PLA, and PP resins. Natural fibers were sometimes used in combination with synthetic fibers to fabricate hybrid composites for structural applications.
- Fabricated composites were subject to various tests to evaluate mechanical characteristics, hygro-thermal degradation, tribological characteristics, electrical characteristics, and ballistic impact performance. The mechanical properties of fibrous composites can be tailored by controlling various factors, such as the fiber orientation, fiber volume fraction, and the matrix type.
- Despite the advantages of fibrous composites, there are still challenges associated with their use, such as manufacturing costs, limited design guidelines, and difficulty in predicting their mechanical behavior.

The field of fibrous composites is continuously evolving, and there are several future directions and challenges that need to be addressed for their widespread use in engineering applications.

- One future direction is to develop sustainable and eco-friendly manufacturing processes for fibrous composites. The use of renewable resources and recyclable materials in the production of composites could reduce their environmental impact and enhance their sustainability.
- Another challenge is to establish standard design guidelines for fibrous composites. The lack of design guidelines hinders their widespread use in industry, as it becomes difficult to determine the optimal fiber-matrix combination, orientation, and thickness for a given application.
- In addition, the development of multi-functional fibrous composites that can provide additional functionalities, such as thermal, electrical, or magnetic properties, would expand their range of applications.

Author Contributions: Writing—original draft preparation, H.S.A. and T.P.J.; writing—review and editing, J.X.; supervision, J.X.; funding acquisition, J.X. All authors have read and agreed to the published version of the manuscript.

Funding: The authors gratefully acknowledge the financial support of the National Natural Science Foundation of China (Grant No. 52175425) and the Shanghai Industrial Collaborative Innovation Project (Grant No. HCXBCY-2022-040). The work was also funded by the 9th Sino-Hungarian Intergovernmental Scientific and Technological Cooperation Project (Grant No. 2021-07).

Data Availability Statement: Not applicable.

Conflicts of Interest: The authors declare no conflict of interest.

Nomenclature

AFs	Aramid Fibers	MWCNTs	Multi-Wall CNTs
AM	Additive Manufacturing	NFRP	Natural Fiber-Reinforced Polymer
BMCs	Bulk Molding Compounds	PA	Polyamide
CFRP	Carbon Fiber-Reinforced Polymer	PLA	Polylactic Acid
CFs	Carbon Fibers	PP	Polypropylene
CM	Compression Molding	PTFE	Polytetrafluoroethylene
CNTs	Carbon Nanotubes	RTM	Resin Transfer Molding
FDM	Fused Deposition Modeling	SCFs	Short Carbon Fibers
FFF	Fused Filament Fabrication	SGFs	Short Glass Fibers
FRPCs	Fiber-Reinforced Polymer Composites	SLA	Stereolithography
FW	Filament Winding	SLS	Selective Laser Sintering
GFs	Glass Fibers	SMCs	Sheet Molding Compounds
GFRP	Glass Fiber-Reinforced Polymer	UHMWPE	Ultra-High-Molecular-Weight Polyethylene
HL	Hand Lay-Up	VARI	Vacuum-Assisted Resin Injection
IM	Injection Molding	VARTM	Vacuum-Assisted Resin Transfer Molding
LDM	Liquid Deposition Modeling	VE	Vinyl Ester
MJF	Multi-Jet Fusion	VI	Vacuum Infusion

References

1. Lunetto, V.; Galati, M.; Settineri, L.; Iuliano, L. Sustainability in the manufacturing of composite materials: A literature review and directions for future research. *J. Manuf. Process.* **2023**, *85*, 858–874. [CrossRef]
2. Xu, J. A review on tool wear issues in drilling CFRP laminates. *Front. Mater.* **2022**, *9*, 990773. [CrossRef]
3. Wang, L.; Zeng, X.; Li, Y.; Yang, H.; Tang, S. Influences of MgO and PVA fiber on the abrasion and cracking resistance, pore structure and fractal features of hydraulic concrete. *Fractal Fract.* **2022**, *6*, 674. [CrossRef]
4. Wang, L.; He, T.; Zhou, Y.; Tang, S.; Tan, J.; Liu, Z.; Su, J. The influence of fiber type and length on the cracking resistance, durability and pore structure of face slab concrete. *Constr. Build. Mater.* **2021**, *282*, 122706. [CrossRef]
5. Divya, H.V.; Jeevan, T.P.; Ashrith, H.S.; Rudresh, B.M. Effect of wood filler on the mechanical and thermal behavior of polypropylene/short glass/short carbon fiber-reinforced hybrid composites. *J. Braz. Soc. Mech. Sci. Eng.* **2022**, *44*, 265. [CrossRef]
6. Xu, J.; Yin, Y.; Davim, J.P.; Li, L.; Ji, M.; Geier, N.; Chen, M. A critical review addressing drilling-induced damage of CFRP composites. *Compos. Struct.* **2022**, *294*, 115594. [CrossRef]
7. Xu, J.; Li, C.; Dang, J.; El Mansori, M.; Ren, F. A study on drilling high-strength CFRP laminates: Frictional heat and cutting temperature. *Materials* **2018**, *11*, 2366. [CrossRef]
8. Khan, H.A.; Nigar, M.; Chaudhry, I.A. Tensile behavior of unidirectional carbon reinforced composites for aerospace structures under varying strain rates. *Appl. Mech. Mater.* **2015**, *798*, 357–361.
9. Amiri, A.; Krosbakken, T.; Schoen, W.; Theisen, D.; Ulven, C.A. Design and manufacturing of a hybrid flax/carbon fiber composite bicycle frame. *Proc. Inst. Mech. Eng. Part P J. Sport. Eng. Technol.* **2018**, *232*, 28–38. [CrossRef]
10. Kim, D.-H.; Kim, H.-G.; Kim, H.-S. Design optimization and manufacture of hybrid glass/carbon fiber reinforced composite bumper beam for automobile vehicle. *Compos. Struct.* **2015**, *131*, 742–752. [CrossRef]
11. Stepanova, M.; Korzhikova-Vlakh, E. Modification of cellulose micro-and nanomaterials to improve properties of aliphatic polyesters/cellulose composites: A review. *Polymers* **2022**, *14*, 1477. [CrossRef]
12. Thakur, V.K.; Thakur, M.K. Processing and characterization of natural cellulose fibers/thermoset polymer composites. *Carbohydr. Polym.* **2014**, *109*, 102–117. [CrossRef] [PubMed]
13. Nguong, C.; Lee, S.; Sujan, D. A review on natural fibre reinforced polymer composites. *Int. J. Mater. Metall. Eng.* **2013**, *7*, 52–59.
14. Vishkaei, M.S.; Salleh, M.A.M.; Yunus, R.; Biak, D.R.A.; Danafar, F.; Mirjalili, F. Effect of short carbon fiber surface treatment on composite properties. *J. Compos. Mater.* **2011**, *45*, 1885–1891. [CrossRef]
15. Botelho, E.; Rezende, M.; Lauke, B. Mechanical behavior of carbon fiber reinforced polyamide composites. *Compos. Sci. Technol.* **2003**, *63*, 1843–1855. [CrossRef]
16. Yan, L.; Chouw, N.; Jayaraman, K. Effect of UV and water spraying on the mechanical properties of flax fabric reinforced polymer composites used for civil engineering applications. *Mater. Des.* **2015**, *71*, 17–25. [CrossRef]
17. Yahaya, R.; Sapuan, S.; Jawaid, M.; Leman, Z.; Zainudin, E. Investigating ballistic impact properties of woven kenaf-aramid hybrid composites. *Fibers Polym.* **2016**, *17*, 275. [CrossRef]
18. Biron, M. *Thermosets and Composites: Material Selection, Applications, Manufacturing and Cost Analysis*; Elsevier: Amsterdam, The Netherlands, 2013.

19. Raji, M.; Abdellaoui, H.; Essabir, H.; Kakou, C.-A.; Bouhfid, R.; Qaiss, A.E.K. Prediction of the cyclic durability of woven-hybrid composites. In *Durability and Life Prediction in Biocomposites, Fibre-Reinforced Composites and Hybrid Composites*; Jawaid, M., Thariq, M., Saba, N., Eds.; Woodhead Publishing: Cambridge, UK, 2019; pp. 27–62.
20. Liu, S.-J. Injection molding in polymer matrix composites. In *Manufacturing Techniques for Polymer Matrix Composites (PMCs)*; Elsevier: Amsterdam, The Netherlands, 2012; pp. 15–46.
21. Kariminejad, M.; Tormey, D.; Huq, S.; Morrison, J.; McAfee, M. Ultrasound sensors for process monitoring in injection moulding. *Sensors* **2021**, *21*, 5193. [CrossRef]
22. Wang, P.; Molimard, J.; Drapier, S.; Vautrin, A.; Minni, J.-C. Monitoring the resin infusion manufacturing process under industrial environment using distributed sensors. *J. Compos. Mater.* **2012**, *46*, 691–706. [CrossRef]
23. Witik, R.A.; Gaille, F.; Teuscher, R.; Ringwald, H.; Michaud, V.; Månson, J.-A.E. Economic and environmental assessment of alternative production methods for composite aircraft components. *J. Clean. Prod.* **2012**, *29*, 91–102. [CrossRef]
24. Yalcinkaya, M.A.; Sozer, E.M.; Altan, M.C. Fabrication of high quality composite laminates by pressurized and heated-VARTM. *Compos. Part A Appl. Sci. Manuf.* **2017**, *102*, 336–346. [CrossRef]
25. Carlone, P.; Baran, I.; Hattel, J.H.; Palazzo, G. Computational approaches for modeling the multiphysics in pultrusion process. *Adv. Mech. Eng.* **2013**, *5*, 301875. [CrossRef]
26. Dai, D.; Fan, M. Wood fibres as reinforcements in natural fibre composites: Structure, properties, processing and applications. In *Natural Fibre Composites*; Elsevier: Amsterdam, The Netherlands, 2014; pp. 3–65.
27. Sentis, D.F.; Cochereau, T.; Orgéas, L.; Dumont, P.J.; Roscoat, S.R.D.; Laurencin, T.; Terrien, M.; Sager, M. Tensile behaviour of uncured sheet moulding compounds: Rheology and flow-induced microstructures. *Compos. Part A Appl. Sci. Manuf.* **2017**, *101*, 459–470. [CrossRef]
28. Nicolais, L.; Kenny, J.M.; Maffezzoli, A.; Torre, L.; Trivisano, A. Science and Technology of Polymer Composites. In *The Interfacial Interactions in Polymeric Composites*; Akovali, G., Ed.; Springer: Dordrecht, The Netherlands, 1993; pp. 321–357.
29. Lee, S.M. *Handbook of Composite Reinforcements*; John Wiley & Sons: Hoboken, NJ, USA, 1996.
30. Li, N.; Li, Y.; Jelonnek, J.; Link, G.; Gao, J. A new process control method for microwave curing of carbon fibre reinforced composites in aerospace applications. *Compos. Part B Eng.* **2017**, *122*, 61–70. [CrossRef]
31. Dhakal, H.N.; Ismail, S.O. Design, manufacturing processes and their effects on bio-composite properties. In *Sustainable Composites for Lightweight Applications*; Dhakal, H.N., Ismail, S.O., Eds.; Woodhead Publishing: Cambridge, UK, 2021; pp. 121–177.
32. Alagirusamy, R. Hybrid yarns for thermoplastic composites. In *Technical Textile Yarns*; Alagirusamy, R., Das, A., Eds.; Woodhead Publishing: Cambridge, UK, 2010; pp. 387–428.
33. Gupta, M.; Jain, A.; Kamineni, J.N.; Burela, R.G. Advances and applications of biofiber-based polymer composites. In *Advances in Bio-Based Fiber*; Rangappa, S.M., Ed.; Woodhead Publishing: Cambridge, UK, 2022; pp. 575–602.
34. Gopanna, A.; Rajan, K.P.; Thomas, S.P.; Chavali, M. Polyethylene and polypropylene matrix composites for biomedical applications. In *Materials for Biomedical Engineering*; Grumezescu, V., Grumezescu, A.M., Eds.; Elsevier: Amsterdam, The Netherlands, 2019; Chapter 6; pp. 175–216.
35. Schäfer, J.; Gries, T. Braiding pultrusion of thermoplastic composites. In *Advances in Braiding Technology*; Kyosev, Y., Ed.; Woodhead Publishing: Cambridge, UK, 2016; pp. 405–428.
36. Ratna, D. Thermoset composites. In *Recent Advances and Applications of Thermoset Resins*, 2nd ed.; Ratna, D., Ed.; Elsevier: Amsterdam, The Netherlands, 2022; Chapter 4; pp. 317–418.
37. Calignano, F.; Manfredi, D.; Ambrosio, E.P.; Biamino, S.; Lombardi, M.; Atzeni, E.; Salmi, A.; Minetola, P.; Iuliano, L.; Fino, P. Overview on additive manufacturing technologies. *Proc. IEEE* **2017**, *105*, 593–612. [CrossRef]
38. Galati, M.; Calignano, F.; Viccica, M.; Iuliano, L. Additive manufacturing redesigning of metallic parts for high precision machines. *Crystals* **2020**, *10*, 161. [CrossRef]
39. Konta, A.A.; García-Piña, M.; Serrano, D.R. Personalised 3D printed medicines: Which techniques and polymers are more successful? *Bioengineering* **2017**, *4*, 79. [CrossRef]
40. Kumar, S.; Kruth, J.-P. Composites by rapid prototyping technology. *Mater. Des.* **2010**, *31*, 850–856. [CrossRef]
41. Goh, G.D.; Yap, Y.L.; Agarwala, S.; Yeong, W.Y. Recent progress in additive manufacturing of fiber reinforced polymer composite. *Adv. Mater. Technol.* **2019**, *4*, 1800271. [CrossRef]
42. Tsouknidas, A. Friction induced wear of rapid prototyping generated materials: A review. *Adv. Tribol.* **2011**, *2011*, 746270. [CrossRef]
43. Matsuzaki, R.; Ueda, M.; Namiki, M.; Jeong, T.-K.; Asahara, H.; Horiguchi, K.; Nakamura, T.; Todoroki, A.; Hirano, Y. Three-dimensional printing of continuous-fiber composites by in-nozzle impregnation. *Sci. Rep.* **2016**, *6*, 23058. [CrossRef]
44. Dharmavarapu, P.; Sreekara Reddy, M.B.S. Aramid fibre as potential reinforcement for polymer matrix composites: A review. *Emergent Mater.* **2022**, *5*, 1561–1578. [CrossRef]
45. Ebrahimnezhad-Khaljiri, H.; Eslami-Farsani, R.; Banaie, K.A. The evaluation of the thermal and mechanical properties of aramid/semi-carbon fibers hybrid composites. *Fibers Polym.* **2017**, *18*, 296–302. [CrossRef]
46. Haque, M.M.; Munshi, M.R.; Alam, S.S.; Haque, M.R.; Hasan, M.; Gafur, M.A.; Ullah, Z.; Sifat, R.S.; Khan, A. Effect of Different Fibers Loading on Palm and Aramid Fiber Reinforced Hybrid Epoxy Composite. *Adv. Mater. Process. Technol.* **2022**, *8*, 1889–1900. [CrossRef]

47. Anidha, S.; Latha, N.; Muthukkumar, M. Reinforcement of Aramid fiber with bagasse epoxy bio-degradable composite: Investigations on mechanical properties and surface morphology. *J. Mater. Res. Technol.* **2019**, *8*, 3198–3212. [CrossRef]
48. Udaya Kumar, P.; Suresha, B.; Rajini, N.; Satyanarayana, K. Effect of treated coir fiber/coconut shell powder and aramid fiber on mechanical properties of vinyl ester. *Polym. Compos.* **2018**, *39*, 4542–4550. [CrossRef]
49. Gürgen, S.; Çelik, O.N.; Kuşhan, M.C. Tribological behavior of UHMWPE matrix composites reinforced with PTFE particles and aramid fibers. *Compos. Part B Eng.* **2019**, *173*, 106949. [CrossRef]
50. Valarmathi, T.; Sangeetha, M.; Venkata, G.G.; Muppala, D.; Siva, R. Hygro-thermal degradation studies on E-glass woven rovings and aramid fiber composites. *Mater. Today Proc.* **2021**, *44*, 3823–3828.
51. Batra, N.; Dikshit, I. Evaluation of mechanical properties of polytherimide reinforced carbon/glass/aramid hybrid composites. *Mater. Today Proc.* **2020**, *33*, 1472–1476. [CrossRef]
52. Arroyo, M.; Bell, M. Morphology/behavior relationship and recyclability of composites based on PP/EPDM blends and short aramid fibers. *J. Appl. Polym. Sci.* **2002**, *83*, 2474–2484. [CrossRef]
53. Müller, K.; Reußmann, T.; Lützkendorf, R.; Schmitt, M. Aramid pulp for the reinforcement of thermoplastic elastomers. *Int. Polym. Sci. Technol.* **2011**, *38*, 162–164. [CrossRef]
54. Louis, B.M.; Maldonado, J.; Klunker, F.; Ermanni, P. Measurement of nanoparticle distribution in composite laminates produced by resin transfer molding. In Proceedings of the 16th European Conference on Composite Materials (ECCM), Seville, Spain, 22–26 June 2014.
55. Song, Y.S. Multiscale fiber-reinforced composites prepared by vacuum-assisted resin transfer molding. *Polym. Compos.* **2007**, *28*, 458–461. [CrossRef]
56. Da Silva, L.F.; Lavoratti, A.; Pereira, I.M.; Dias, R.R.; Amico, S.C.; Zattera, A.J. Development of multilaminar composites for vehicular ballistic protection using ultra-high molecular weight polyethylene laminates and aramid fabrics. *J. Compos. Mater.* **2019**, *53*, 1907–1916. [CrossRef]
57. Ding, X.; Zhang, Z.; Kong, H.; Qiao, M.; Hu, Z.; Zhang, L.; Yu, M. Influences of graphene oxide addition on mechanical properties of aramid fiber reinforced composites. *Mater. Express* **2019**, *9*, 578–586. [CrossRef]
58. Nagendra, J.; Prasad, M.G. FDM process parameter optimization by Taguchi technique for augmenting the mechanical properties of nylon–aramid composite used as filament material. *J. Inst. Eng. Ser. C* **2020**, *101*, 313–322. [CrossRef]
59. Chen, J.; Tan, P.; Liu, X.; Tey, W.S.; Ong, A.; Zhao, L.; Zhou, K. High-strength light-weight aramid fibre/polyamide 12 composites printed by Multi Jet Fusion. *Virtual Phys. Prototyp.* **2022**, *17*, 295–307. [CrossRef]
60. Xu, J.; Li, C.; Mansori, M.E.; Liu, G.; Chen, M. Study on the Frictional Heat at Tool-Work Interface when Drilling CFRP Composites. *Procedia Manuf.* **2018**, *26*, 415–423. [CrossRef]
61. Geier, N.; Xu, J.; Pereszlai, C.; Poór, D.I.; Davim, J.P. Drilling of carbon fibre reinforced polymer (CFRP) composites: Difficulties, challenges and expectations. *Procedia Manuf.* **2021**, *54*, 284–289. [CrossRef]
62. Altin Karataş, M.; Gökkaya, H. A review on machinability of carbon fiber reinforced polymer (CFRP) and glass fiber reinforced polymer (GFRP) composite materials. *Def. Technol.* **2018**, *14*, 318–326. [CrossRef]
63. Geier, N.; Davim, J.P.; Szalay, T. Advanced cutting tools and technologies for drilling carbon fibre reinforced polymer (CFRP) composites: A review. *Compos. Part A Appl. Sci. Manuf.* **2019**, *125*, 105552. [CrossRef]
64. Zhou, Y.; Zhang, P.; Ning, F. Joining of carbon fiber reinforced polymer/titanium stacks using directed energy deposition additive manufacturing. *Compos. Struct.* **2023**, *310*, 116775. [CrossRef]
65. Ji, M.; Xu, J.; Chen, M.; Mansori, M.E.I. Effects of Different Cooling Methods on the Specific Energy Consumption when Drilling CFRP/Ti6Al4V Stacks. *Procedia Manuf.* **2020**, *43*, 95–102. [CrossRef]
66. Xu, J.; Ji, M.; Chen, M.; El Mansori, M. Experimental investigation on drilling machinability and hole quality of CFRP/Ti6Al4V stacks under different cooling conditions. *Int. J. Adv. Manuf. Technol.* **2020**, *109*, 1527–1539. [CrossRef]
67. Badakhsh, A.; An, K.-H.; Kim, B.-J. Enhanced surface energetics of CNT-grafted carbon fibers for superior electrical and mechanical properties in CFRPs. *Polymers* **2020**, *12*, 1432. [CrossRef] [PubMed]
68. Srikanth, I.; Kumar, S.; Kumar, A.; Ghosal, P.; Subrahmanyam, C. Effect of amino functionalized MWCNT on the crosslink density, fracture toughness of epoxy and mechanical properties of carbon–epoxy composites. *Compos. Part A: Appl. Sci. Manuf.* **2012**, *43*, 2083–2086. [CrossRef]
69. Eslami, Z.; Yazdani, F.; Mirzapour, M.A. Thermal and mechanical properties of phenolic-based composites reinforced by carbon fibres and multiwall carbon nanotubes. *Compos. Part A Appl. Sci. Manuf.* **2015**, *72*, 22–31. [CrossRef]
70. Tariq, F.; Shifa, M.; Baloch, R.A. Mechanical and thermal properties of multi-scale carbon nanotubes–carbon fiber–epoxy composite. *Arab. J. Sci. Eng.* **2018**, *43*, 5937–5948. [CrossRef]
71. Poodts, E.; Minak, G.; Mazzocchetti, L.; Giorgini, L. Fabrication, process simulation and testing of a thick CFRP component using the RTM process. *Compos. Part B: Eng.* **2014**, *56*, 673–680. [CrossRef]
72. Jagannatha, T.; Harish, G. Mechanical properties of carbon/glass fiber reinforced epoxy hybrid polymer composites. *Int. J. Mech. Eng. Robot. Res.* **2015**, *4*, 131–137.
73. Fu, S.-Y.; Lauke, B.; Mäder, E.; Yue, C.-Y.; Hu, X.; Mai, Y.-W. Hybrid effects on tensile properties of hybrid short-glass-fiber-and short-carbon-fiber-reinforced polypropylene composites. *J. Mater. Sci.* **2001**, *36*, 1243–1251. [CrossRef]
74. Junaedi, H.; Albahkali, E.; Baig, M.; Dawood, A.; Almajid, A. Ductile to brittle transition of short carbon fiber-reinforced polypropylene composites. *Adv. Polym. Technol.* **2020**, *2020*, 6714097. [CrossRef]

75. Gupta, A.; Fidan, I.; Hasanov, S.; Nasirov, A. Processing, mechanical characterization, and micrography of 3D-printed short carbon fiber reinforced polycarbonate polymer matrix composite material. *Int. J. Adv. Manuf. Technol.* **2020**, *107*, 3185–3205. [CrossRef]
76. Dickson, A.N.; Barry, J.N.; McDonnell, K.A.; Dowling, D.P. Fabrication of continuous carbon, glass and Kevlar fibre reinforced polymer composites using additive manufacturing. *Addit. Manuf.* **2017**, *16*, 146–152. [CrossRef]
77. Yan, C.; Hao, L.; Xu, L.; Shi, Y. Preparation, characterisation and processing of carbon fibre/polyamide-12 composites for selective laser sintering. *Compos. Sci. Technol.* **2011**, *71*, 1834–1841. [CrossRef]
78. Wang, L.; Kiziltas, A.; Mielewski, D.F.; Lee, E.C.; Gardner, D.J. Closed-loop recycling of polyamide12 powder from selective laser sintering into sustainable composites. *J. Clean. Prod.* **2018**, *195*, 765–772. [CrossRef]
79. Rajak, D.K.; Wagh, P.H.; Linul, E. Manufacturing technologies of carbon/glass fiber-reinforced polymer composites and their properties: A review. *Polymers* **2021**, *13*, 3721. [CrossRef] [PubMed]
80. Tsao, V. Experimental study of drilling composite materials with step-core drill. *Mater. Des.* **2008**, *29*, 1740–1744. [CrossRef]
81. Palanikumar, K. Experimental investigation and optimisation in drilling of GFRP composites. *Measurement* **2011**, *44*, 2138–2148. [CrossRef]
82. Choudhary, M.; Sharma, A.; Dwivedi, M.; Patnaik, A. A comparative study of the physical, mechanical and thermo-mechanical behavior of GFRP composite based on fabrication techniques. *Fibers Polym.* **2019**, *20*, 823–831. [CrossRef]
83. Kim, S.-Y.; Shim, C.S.; Sturtevant, C.; Kim, D.; Song, H.C. Mechanical properties and production quality of hand-layup and vacuum infusion processed hybrid composite materials for GFRP marine structures. *Int. J. Nav. Archit. Ocean. Eng.* **2014**, *6*, 723–736. [CrossRef]
84. Faudree, M.C.; Nishi, Y.; Salvia, M. Increasing impact strength of a short glass fiber compression molded BMC by shortening fibers without change in equipment. *Materials* **2022**, *15*, 1145. [CrossRef] [PubMed]
85. Khalid, M.Y.; Arif, Z.U.; Sheikh, M.F.; Nasir, M.A. Mechanical characterization of glass and jute fiber-based hybrid composites fabricated through compression molding technique. *Int. J. Mater. Form.* **2021**, *14*, 1085–1095. [CrossRef]
86. Rafiee, R. Experimental and theoretical investigations on the failure of filament wound GRP pipes. *Compos. Part B: Eng.* **2013**, *45*, 257–267. [CrossRef]
87. Ahmed, K.S.; Vijayarangan, S. Tensile, flexural and interlaminar shear properties of woven jute and jute-glass fabric reinforced polyester composites. *J. Mater. Process. Technol.* **2008**, *207*, 330–335. [CrossRef]
88. Huang, G.; Sun, H. Effect of water absorption on the mechanical properties of glass/polyester composites. *Mater. Des.* **2007**, *28*, 1647–1650. [CrossRef]
89. Lee, G.-W.; Lee, N.-J.; Jang, J.; Lee, K.-J.; Nam, J.-D. Effects of surface modification on the resin-transfer moulding (RTM) of glass-fibre/unsaturated-polyester composites. *Compos. Sci. Technol.* **2002**, *62*, 9–16. [CrossRef]
90. Sano, Y.; Matsuzaki, R.; Ueda, M.; Todoroki, A.; Hirano, Y. 3D printing of discontinuous and continuous fibre composites using stereolithography. *Addit. Manuf.* **2018**, *24*, 521–527. [CrossRef]
91. Li, Z.; Zhou, W.; Yang, L.; Chen, P.; Yan, C.; Cai, C.; Li, H.; Li, L.; Shi, Y. Glass fiber-reinforced phenol formaldehyde resin-based electrical insulating composites fabricated by selective laser sintering. *Polymers* **2019**, *11*, 135. [CrossRef]
92. Benal, M.G.M.; GS, P.K.; Tambrallimath, V.; HR, G.; Khan, T.Y.; Rajhi, A.A.; Baig, M.A.A. Influence of Short Glass Fibre Reinforcement on Mechanical Properties of 3D Printed ABS-Based Polymer Composites. *Polymers* **2022**, *14*, 1182.
93. Sai Shravan Kumar, P.; Viswanath Allamraju, K. A Review Of Natural Fiber Composites [Jute, Sisal, Kenaf]. *Mater. Today Proc.* **2019**, *18*, 2556–2562. [CrossRef]
94. Li, M.; Pu, Y.; Thomas, V.M.; Yoo, C.G.; Ozcan, S.; Deng, Y.; Nelson, K.; Ragauskas, A.J. Recent advancements of plant-based natural fiber–reinforced composites and their applications. *Compos. Part B Eng.* **2020**, *200*, 108254. [CrossRef]
95. Wilson, A. Vehicle weight is the key driver for automotive composites. *Reinf. Plast.* **2017**, *61*, 100–102. [CrossRef]
96. Huda, M.S.; Drzal, L.T.; Ray, D.; Mohanty, A.K.; Mishra, M. Natural-fiber composites in the automotive sector. In *Properties and Performance of Natural-Fibre Composites*; Pickering, K.L., Ed.; Woodhead Publishing: Cambridge, UK, 2008; pp. 221–268.
97. Saravana Kumar, A.; Selvi, P.M.; Rajeshkumar, L. Delamination in drilling of sisal/banana reinforced composites produced by hand lay-up process. *Appl. Mech. Mater.* **2017**, *867*, 29–33.
98. Meenakshi, C.; Krishnamoorthy, A. Preparation and mechanical characterization of flax and glass fiber reinforced polyester hybrid composite laminate by hand lay-up method. *Mater. Today Proc.* **2018**, *5*, 26934–26940. [CrossRef]
99. Seetha Girish, A.V.; Vijaya, B.; Arjun, S.; Raghavendra, S. Manufacturing and mechanical characterization of coir fibre composites based on vinyl ester. *Adv. Mater. Process. Technol.* **2022**, *8*, 1997–2006.
100. Prasad, V.; Hunize, C.M.; Abhiraj, R.; Jospeh, M.; Sekar, K.; Ali, M. Mechanical properties of flax fiber reinforced composites manufactured using hand layup and compression molding—A comparison. In *Advances in Industrial and Production Engineering: Select Proceedings of FLAME 2018*; Springer: Berlin/Heidelberg, Germany, 2019.
101. Zhang, J.; Khatibi, A.A.; Castanet, E.; Baum, T.; Komeily-Nia, Z.; Vroman, P.; Wang, X. Effect of natural fibre reinforcement on the sound and vibration damping properties of bio-composites compression moulded by nonwoven mats. *Compos. Commun.* **2019**, *13*, 12–17. [CrossRef]
102. Oksman, K. High quality flax fibre composites manufactured by the resin transfer moulding process. *J. Reinf. Plast. Compos.* **2001**, *20*, 621–627. [CrossRef]

103. Gorrepotu, S.R.; Debnath, K.; Mahapatra, R.N. Multi-response Optimization of the Chemical Treatment Process Parameters Influencing the Tensile, Flexural, Compression, and Shear Properties of the Injection Moulded Green Composites. *J. Polym Environ.* **2023**, *31*, 112–130. [CrossRef]
104. Long, H.; Wu, Z.; Dong, Q.; Shen, Y.; Zhou, W.; Luo, Y.; Zhang, C.; Dong, X. Mechanical and thermal properties of bamboo fiber reinforced polypropylene/polylactic acid composites for 3D printing. *Polym. Eng. Sci.* **2019**, *59*, E247–E260. [CrossRef]
105. Anandkumar, R.; Babu, S.R.; Sathyamurthy, R. Investigations on the mechanical properties of natural fiber granulated composite using hybrid additive manufacturing: A novel approach. *Adv. Mater. Sci. Eng.* **2021**, *2021*, 5536171. [CrossRef]
106. Zhao, D.; Guo, Y.; Jiang, K.; Zhang, H. Preparation and selective laser sintering of bamboo flour/copolyester composite and post-processing. *J. Thermoplast. Compos. Mater.* **2017**, *30*, 1045–1055. [CrossRef]
107. Jaiswal, G.; Singha, M.K.; Das, D. Mechanical behavior of aramid-polypropylene fiberweb composites. *Compos. Struct.* **2021**, *268*, 113938. [CrossRef]
108. Kandekar, S.; Talikoti, R. Study of torsional behavior of reinforced concrete beams strengthened with aramid fiber strips. *Int. J Adv. Struct. Eng.* **2018**, *10*, 465–474. [CrossRef]
109. Liu, X.; Tey, W.S.; Tan, P.; Leong, K.K.; Chen, J.; Tian, Y.; Ong, A.; Zhao, L.; Zhou, K. Effect of the fibre length on the mechanical anisotropy of glass fibre–reinforced polymer composites printed by Multi Jet Fusion. *Virtual Phys. Prototyp.* **2022**, *17*, 734–748. [CrossRef]
110. Thomason, J. Dependence of interfacial strength on the anisotropic fiber properties of jute reinforced composites. *Polym. Compos* **2010**, *31*, 1525–1534. [CrossRef]
111. Ouali, A.-A.; Rinberg, R.; Nendel, W.; Kroll, L.; Siegel, C.; Buchelt, B.; Wagenführ, A.; Trommler, K.; Schreiter, K.; John, R. Natural unidirectional sheet processes for fibre reinforced bioplastics. *AIP Conf. Proc.* **2017**, *1914*, 060005.
112. Couture, A.; Lebrun, G.; Laperrière, L. Mechanical properties of polylactic acid (PLA) composites reinforced with unidirectional flax and flax-paper layers. *Compos. Struct.* **2016**, *154*, 286–295. [CrossRef]

Journal of
Composites Science

MDPI

Review

Cool-Clave—An Energy Efficient Autoclave

Indraneel R. Chowdhury and John Summerscales *

School of Engineering, Computing and Mathematics, University of Plymouth, Drake Circus, Plymouth PL4 8AA, UK
* Correspondence: j.summerscales@plymouth.ac.uk

Abstract: Out-of-autoclave (OOA) manufacturing techniques for composites result in lower fibre volume fractions than for fully compressed laminates. The lower fibre volume fraction produces a higher resin volume fraction, which becomes resin-rich volumes (RRV). Textile reinforcements with clustered fibres and consequent RRV generally have low strength but high in-plane process permeability, whereas the opposite is true for uniformly distributed fibres. The inevitable increase in resin volume fraction of OOA composites often compromises composite performance and leads to relatively higher weight and fuel consumption in transport applications. The retention of autoclave processing is recommended for highest performance when compression press moulding is not appropriate (for example, for complex 3D components). The traditional autoclave processing of composites heats not only the component to be cured but also parasitic air and the vessel insulation. Subject to minor modifications of the pressure vessel, electrically heated tooling could be implemented. This approach would need to balance insulation of the heated tool surface (and any heater blanket on the counter-face) against the quenching effect during the introduction of the pressurised cool air. This process optimisation would significantly reduce energy consumption. Additionally, the laminate on the heated tool could be taken to the end of the dwell period before loading the autoclave, leading to significant reductions in cure cycle times. Components could be cured simultaneously at different temperatures provided that there are sufficient power and control circuits in the autoclave. While autoclave processing has usually involved vacuum-bagged pre-impregnated reinforcements, implementation of the cool-clave technique could also provide a scope for using the pressure vessel to cure vacuum-infused composites.

Keywords: autoclave; cool-clave; vacuum; heated tooling; fibre-reinforced composites

Citation: Chowdhury, I.R.; Summerscales, J. Cool-Clave—An Energy Efficient Autoclave. *J. Compos. Sci.* **2023**, *7*, 82. https://doi.org/10.3390/jcs7020082

Academic Editor: Jinyang Xu

Received: 23 January 2023
Revised: 8 February 2023
Accepted: 14 February 2023
Published: 16 February 2023

1. Introduction

Composite materials are widely used in industrial applications due to their unique characteristics of high-stiffness-to-weight ratio, excellent durability, chemical resistance, and better recycling potential as compared to metallic components. High performance composites, mainly used in aerospace applications, are produced in the autoclave by applying elevated pressure and temperature [1,2]. However, autoclave processing of composites involves long curing cycle times, expensive tooling, and high energy consumption [3,4]. As a result, there is an interest across the range of composites manufacturing processes for cost reduction with a current focus on out-of-autoclave (OOA) processes [1,5], especially OOA prepreg [6] and resin infusion under flexible tooling [7–9]. The OOA process involves manufacturing composites by applying vacuum and heat outside of the autoclave, but has limitations on the maximum laminate fibre volume fraction due to compressibility characteristics of the reinforcement. As a result, composites manufactured by vacuum-only processes cannot achieve high fibre volume contents, which is a primary requirement in high performance composites for aerospace, automobile, and defence sectors. Compression moulding in a hydraulic press creates limited compaction perpendicular to the line of action of the press. The autoclave is the best process for consolidation of complex three-dimensional components, but suffers from several limitations: (a) pre-impregnated (prepreg) reinforcements incur a high-cost premium, (b) high energy input for heating and air circulation: flow speeds in the range 1.4–2.1 m/s [5], (c) non-uniform heating of the components in the vessel due to turbulent flow, windward vs. leeward location, flow stagnation and consequent temperature differences, (d) thermal lag due to the tool or consumables between the heat source and the composite, and (e) long cycle times, which may be a "bottleneck" constraint [10,11].

However, with minor modifications of the pressure vessel, there is potential to improve the efficiency of autoclave processes. The traditional autoclave not only heats the composite parts to be cured but also inert gasses and vessel insulation. Significant energy savings can result from using electrically heated tooling to only heat the essential parts of the process (the tool and composite), and by cool-air pressurisation of composites [12]. Further to the potential for significant reductions in energy consumption, the laminate on the heated tool could be taken to the end of the dwell period before loading the autoclave, leading to significant reductions in cycle times. Autoclave loading efficiency could be improved by curing different composite systems simultaneously with the composites brought to their respective curing temperatures before loading the autoclave, provided there are sufficient power and control circuits in the autoclave, which would further enhance process efficiency [12].

This paper critically reviews the technique of using heated tooling in the autoclave to enhance the energy- and cost-efficiency of autoclave process, designated as the 'cool-clave' technique.

2. Autoclave

Autoclaves have become indispensable equipment to process high-quality polymer composite materials for structural industries, such as aerospace, automotive, and defence sectors [13]. Today, for example, in the aircraft industry, investments in such equipment are strategically important. Autoclaves are now being used to produce very large aircraft components, such as wing and fuselage sections. They can process a wide variety of materials, including thermoset [14] and thermoplastic [15]-based composite parts with varying contours and complex shapes.

The quality requirements of the present high-performance composites for aerospace/defence industries are indeed more stringent. Additionally, there is an urgent requirement to improve the efficiency and cost-effectiveness of high-performance structural composites, along with ensuring reliable and consistent processing methods. Therefore, it is imperative for autoclave design engineers to take into consideration different governing criteria to address the diverse and complex requirements for developing state-of-the-art autoclave systems. In addition to handling a wide variety of consumables, modern autoclaves must respect health and safety requirements [16] and ensure minimum maintenance costs.

Autoclaves are closed pressure vessels used to manufacture high performance composite components. Uncured composites are moulded in an autoclave typically heated using inert gases, such as carbon dioxide or nitrogen, thus allowing the transfer of heat and pressure to the composite component for consolidation and allowing it to cure firmly and uniformly. The application of pressure for consolidation of composites in an autoclave helps in reducing porosity and voids, retains shape around the mould, and enables better control to maintain a higher fibre volume fraction in composite components [17]. The autoclave process draws many similarities with hot pressing technique; however, the main difference pertains to the way heat and pressure are applied [18]. The autoclave operating parameters, such as temperature and pressure are based on the resin systems used. Generally, epoxy resins require temperatures < 200 °C and pressures of 0.7 MPa [17].

Figure 1 shows the internal chamber of the Aeroform autoclave located in the composites manufacturing laboratory at the University of Plymouth. Surrounding the main internal area is a metal inner case which shields the components being cured from twelve electric heating elements positioned at intervals around the chamber's circumference. Behind the heating elements is a layer of thermal insulation, protected by sheet metal. As the autoclave walls are made up of quality carbon steel, up to 150 mm on some autoclaves [17,18], they act as a heat sink. The insulation is designed to minimise heat transfer from the main chamber to the autoclave walls. The insulation commonly used consists of refractory ceramics or fibreglass insulation. The insulation prevents excess energy loss and is designed to keep the autoclave outer walls down to a maximum temperature of 60 °C.

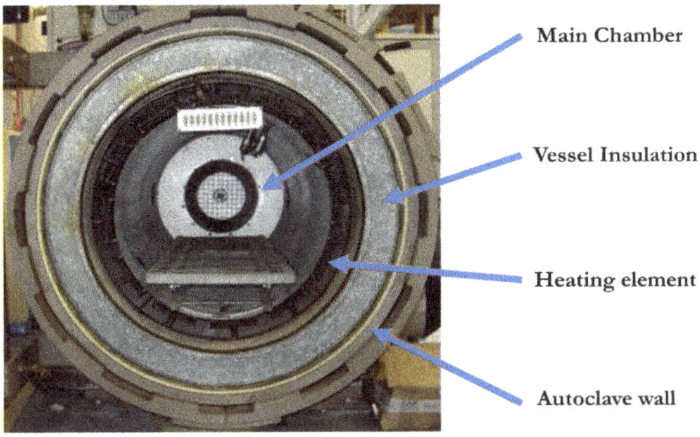

Figure 1. Internal chamber of 0.67 m working diameter Aeroform autoclave at the University of Plymouth.

The design of autoclave systems is multidisciplinary in nature and encompasses mechanical, process control, and instrumentation engineering. Invariably, the state-of-the-art autoclave systems are completely computer-controlled and are semi- or fully automated. The computer controls of these modern autoclaves are required to execute the selected cure cycle by sequentially starting various subsystems, download set values at regular time intervals to the front-end controllers, acquire, store, and archive the data, monitor cure status and faults, generate alarms, and perform the functions of sequential shut down and reporting [19]. Ease of maintenance, fail-safe operation, and reliability are among the key drivers in modern autoclaves. The low cost of ownership also needs to be considered in today's context [17]. In recent years, there has been an increasing demand to enhance the service temperature of high performance structural composite components, invariably leading to higher curing temperature and pressure requirements in the order of 300 to 350 °C and up to 1.5 MPa. This necessitates the development of high temperature and high-pressure autoclave systems, which presents a new set of challenges such as the handling of massive door and locking systems, temperature uniformity, special material requirements for door and shell flanges, fabrication, transportation, and, most importantly, low cost and maintenance requirements [17,19].

2.1. Autoclave Moulding

In an autoclave moulding, isostatic pressure is applied to the composite component on a vacuum bagged mould prior to applying heat/vacuum/pressure to compact the composite material. The intimacy of contact between the composite and mould therefore depends on the magnitude of the applied pressure. Pre-impregnated (prepreg) reinforcements cured by autoclave processing are first covered in peel-ply, with bleeders used to draw out excess air and to soak up excess resin while curing (unless the system is intended for zero-bleed), prior to vacuum bagging, as this allows for the manufacturing of composite components with a high-quality surface finish. During autoclave processing, engineers adjust the process by maximising air expulsion and minimising excess resin flow. Curing pressures typically ranges between 3–12 MPa [20,21]. Prepreg reinforcements are generally used as they offer good processability. The fundamental components of an autoclave moulding process are shown in Figure 2.

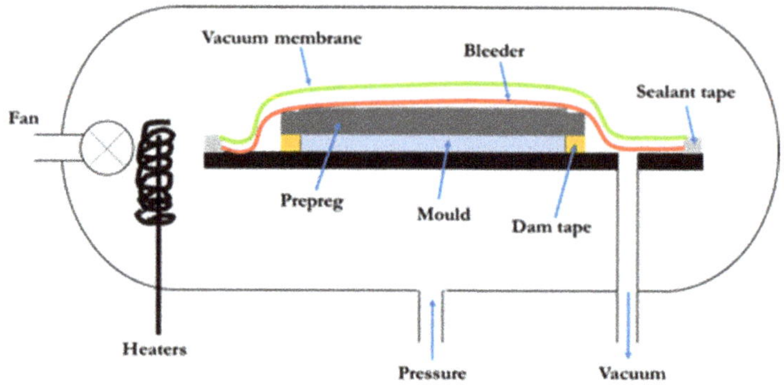

Figure 2. Fundamental components of an Autoclave process (redrawn from [20]).

Autoclave manufacturing is generally a 3-stage process. In the first stage, the vacuum is applied until the desired temperature is reached for composite consolidation. The ramped heating enables resin viscosity to gradually decrease, as well as allowing for the release of volatiles and air bubbles. The reduced viscosity of the resin simultaneously improves wetting of the fibres and resin flow, which in turn facilities movements of volatiles and air bubbles. In the second stage, after a suitable dwell period, the consolidation pressure is applied. The final stage raises temperature to post-cure the composite, following which the resin viscosity stabilises and the material starts to cure [20]. A typical autoclave moulding bagging mechanism is further illustrated in Figure 3.

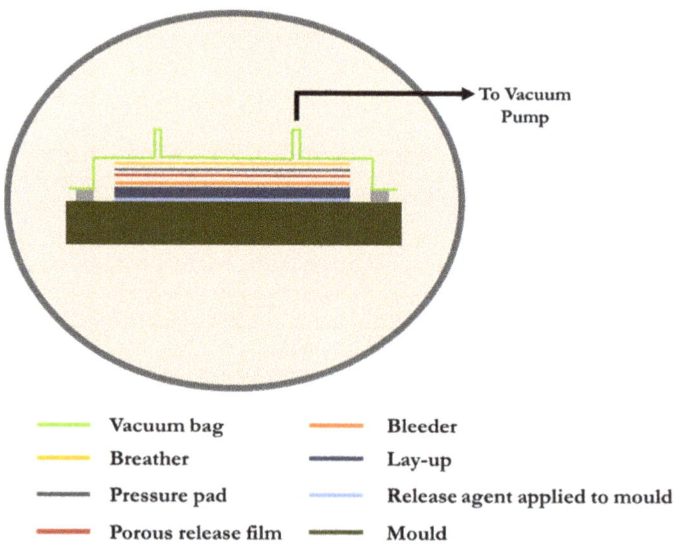

Figure 3. A typical autoclave moulding bagging mechanism.

In thermoset resins, curing is a chemical cross-linking process that is very crucial for optimum performance of thermoset matrix composites. For thermoplastic matrix composites, autoclave consolidation is achieved at high temperatures and bagging materials, and techniques used for thermoset composites are adapted to suit thermoplastics [22].

Autoclave moulding process comprises the following advantages [23]:

(a) Applicable to both thermosetting and thermoplastic fibre-reinforced composites
(b) Facilitates better inter-layer adhesion
(c) Enables good fibre wetting
(d) Enables higher uniformity in composite solidification
(e) Supports superior fibre volume fraction in composite component
(f) Reduced void content in the composite due to the vacuum/pressure
(g) Facilitates manufacturing composite parts with high strength-to-weight ratio

The benefits of the autoclave are that the heat and pressure deliver high performance components for the composite industry, to compete with and replace metal components. Therefore, lightweight components, for example composite motor vehicle parts, can achieve higher efficiencies. In this scenario, engines will require less torque and fuel than when used to move a heavy metal structure. The disadvantages include high capital cost of the machine, and non-flexible and poor heat transfer efficiency. To heat the composite component, it is normal to parasitically heat all the air in the pressure vessel and the thermal insulation of the vessel. Most of the energy that is put into the system is taken up by the heater, therefore creating a higher consumption of energy. Other disadvantages include higher waiting times due to slow ramp rates and longer curing cycles [24].

2.2. Heating and Air-Circulation System in an Autoclave

Forced gas circulation systems using nitrogen or carbon dioxide are most commonly used in autoclaves. The air circulation system consists of a centrifugal blower and ducting system. The heating elements are placed around the impeller. The centrifugal blower takes in gas axially and discharges it radially to pass over the heating elements at a velocity of 1–2 m/s at ambient conditions. The air circulation system also helps in accelerating the cooling process by removing the gas from the outer surface of the cooling tubes at an increasing rate. Modern autoclaves are mechanised with a flange-mounted blower motor encased within a pressure-tight casing and connected to the rear of the autoclave. This enables the motor rotor, stator, and the mechanical components, for example, bearings to directly encounter the autoclave pressure. Power ratings of a typical autoclave range between 100 to 150 kW [17].

Heating system in autoclaves are either electrically controlled or controlled by indirect gas firing (circulating externally heated or cooled thermic fluid). However, the majority of the autoclaves are electrically heated, as these systems are cleaner and more compatible to modern computer control systems and provide better control of autoclave temperature. The electrical heating requirement in an autoclave is based on the charge and resin system requirements for the cure cycle. For example, a typical 4.5-m diameter × 9-m length autoclave requires a heating capacity of approximately 1 MW. Heating elements, typically ranging between 5–10 kW, are usually manufactured using nichrome/kanthal filament with an outer sheath of steel grouped together in banks and connected in star or delta configuration [17]. The air circulation and heating system of a typical autoclave is illustrated in Figure 4.

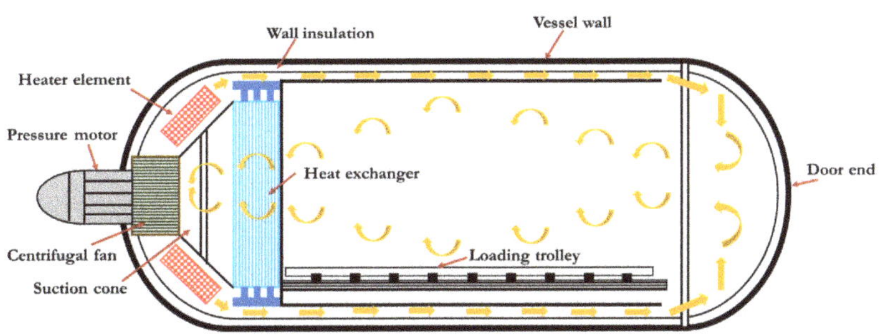

Figure 4. Schematic of air circulation and heating system in a typical autoclave (redrawn from [17]).

2.3. Pressurization and Cooling System

The pressurization systems in an autoclave ensure that the required pressurization rate is maintained during autoclave processing of composites. The average pressurization rate in modern autoclaves is 0.2 MPa/min [17]. Many modern autoclaves are nitrogen-pressurized instead of air-pressurized since autoclave cure consumables are highly inflammable in air medium due to the presence of oxygen. The pressurization system in an autoclave consists of a primary compressor, booster compressor, storage tanks, and piping circuitry. The primary compressor takes in air from the atmosphere and pressurizes it to 0.7 MPa The booster compressor further pressurizes the air to high pressure (typically in the range between 1.7–2.2 MPa) in order to create sufficient pressure differentials to attain the required pressurization rate. In nitrogen-pressurized autoclaves, the nitrogen plant receives the air from the primary compressor at 0.7 MPa pressure and isolates nitrogen from atmospheric air by a process called Pressure Swing Adsorption (PSA) [17], which can produce nitrogen in the order of 99% purity suitable for curing polymeric composite materials. Nitrogen is then pressurized by the booster compressor to meet the required pressurization rate and purged into the chamber [17]. In Figure 5, the cooling and pressurization system used in modern autoclaves is schematically represented.

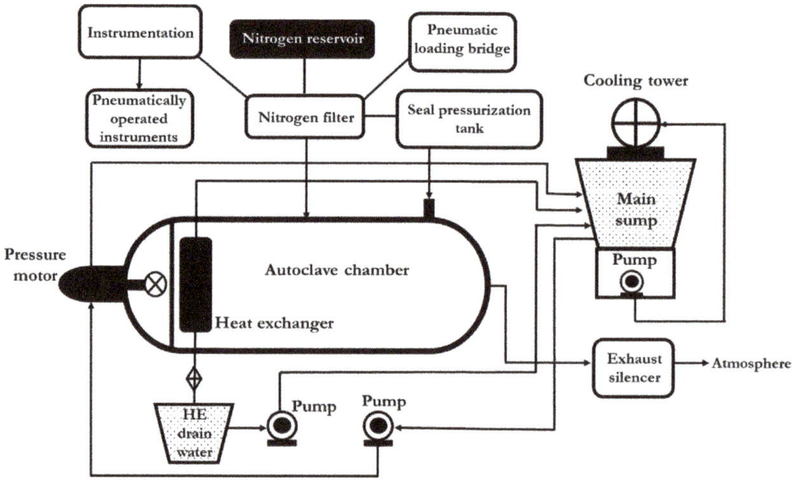

Figure 5. Schematic of cooling and pressurization system in an autoclave (redrawn from [17]).

Autoclave processing requires variable cooling rates based on the resin system used. Other variables affecting the cooling system are (a) temperature difference between the autoclave ambient and cooling medium, (b) flow rate of cooling medium, (c) area of heat transfer, (d) type of flow of cooling medium—parallel or cross flow, (e) conductivity of cooling coil material, and (f) velocity of autoclave medium across heat exchanger. The cooling rate in an autoclave is controlled by varying the flow of cooling fluid. In some autoclaves, both air and water are used as the cooling medium. In autoclaves, a closed loop cooling system is generally employed to prevent excess use of water. Liquid nitrogen is also used for faster cooling. A major challenge in designing cooling system for autoclaves is how fast and effectively the cooling medium can be drained from the autoclave heat exchanger, as any delay in draining the cooling medium from heat exchanger can lead to loss of heat during the heating phase and damage to the heat exchanger tubes. In the water-based cooling system of autoclaves, the simplest way of draining the water is to provide a sump just below the autoclave heat exchanger and then pump back the water to the cooling water sump. This process helps in reducing wastage of water and at the same time also prevents the steam from entering the cooling tower [17].

2.4. Vacuum System

Figure 6 illustrates the vacuum system of a typical autoclave. The main components of vacuum system in modern autoclaves are vacuum pumps, vacuum reservoirs, buffer tanks, suction, and measurement lines. All modern autoclaves must include an adequate number of vacuum ports and must also have the capacity to maintain different levels of vacuum in different bagging systems simultaneously. Therefore, vacuum pumps and reservoirs in an autoclave must have adequate buffer capacity. For example, vacuum pump with minimum capacity of 7 m^3/h is necessary for a bagging area of approximately 1 m^2. Generally, a 4.5 m × 9 m autoclave can have approximately 60 measurement and suction lines. Correspondingly, the pump capacity ranges to about 180 m^3/h with a reservoir capacity of 6 m^3. Vacuum requirement ranges between 667–26,664 × 10^{-6} MPa based on the curing system [17].

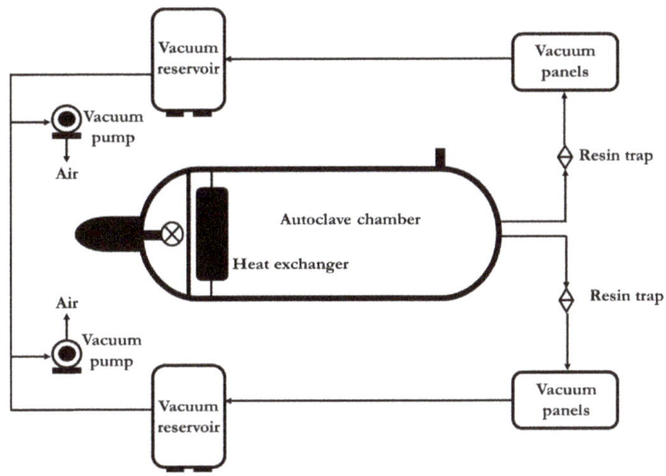

Figure 6. Schematic of vacuum system in a typical autoclave (redrawn from [17]).

2.5. Loading System

The loading system in an autoclave facilitates positioning of composite components or moulds to be cured. A loading crab was used to move the loading platform in an autoclave. Generally, autoclaves are installed in a pit with the top surface of the loading platform in flush with the floor for convenience in loading components. A loading bridge is generally deployed to bridge the gap between autoclave door and pit.

2.6. Electrical and Control System

Electrical and control systems in autoclaves play a key role in ensuring safe operation for reliable processing and curing of composite components. The electrical and control system of autoclaves should be robust enough to be able to provide necessary feedback signals and respond to various commands for processing [25]. Autoclaves are generally built with computer system, serial port servers, power supplies, sensors, and are generally resilient enough to be able to operate even if one or more components fail. Electrical and control systems in autoclaves are in-built with capabilities to be operated in multiple modes of operations, for examples, automatic, semi-automatic, or in manually operational mode. The control system generally consists of PID (proportional integral derivative) controllers, set temperature, and vacuum and pressure levels. PLC (programmable logic controllers) are used to ensure safe interlocking, sequential operation, status, and alarm display. All the components are generally connected to a server and computer-controlled via ethernet links. The communication system used in autoclaves includes RS485, USB, and ethernet connection [26].

3. Cool-Clave Technique

Autoclave processing is used to manufacture high performance composites, but over all, the process is very expensive and requires high energy consumption. In a traditional autoclave, the process not only heats up the composite to be consolidated but also any parasitic air and vessel insulation requiring high energy usage. Additionally, autoclave processing is normally used to consolidate vacuum-bagged pre-impregnated reinforcements. Prepreg materials require freezer storage, and the separate impregnation stage incurs further costs. The cool-clave technique has the potential to achieve autoclave consolidation of fibre-reinforced composites by enhancing the energy- and cost-efficiency: (a) without heating parasitic materials (hence saving energy), (b) with shorter cycle times (hence increased production rate), (c) without prepreg, i.e., using infused laminates (for lower cost). A schematic of the cool-clave technique with the bagging process is illustrated in Figure 7.

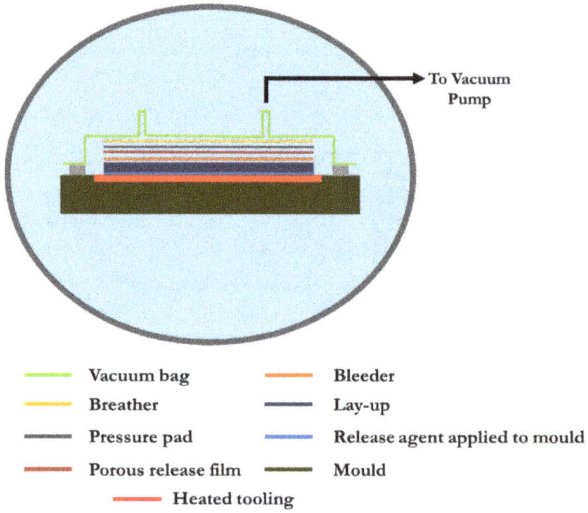

Figure 7. Schematic of cool-clave processing technique.

Implementation of the cool-clave technique for autoclave consolidation of composites could generate significant cost and energy savings. The equipment required for the cool-clave process consists of:

- pressure vessel (autoclave)
- heated mould tool and control unit
- efficient thermal insulation of the bagged component

Prudham and Summerscales [27] investigated the cool-clave processing technique using the autoclave at the University of Plymouth. The autoclave was used as a pressure vessel, without using the heating aspect of the system. The heat, required for curing, was provided by a mould tool with built-in heating elements. Figure 8 demonstrates the heated mould tool used for the cool-clave processing.

For implementation of the cool-clave processing technique, temperature uniformity across the face of the mould tool is crucial [18]. During the initial stages of the cure cycle, dwell periods are often used, where the laminate is held at a temperature lower than the curing temperature for a set period of time. This increase in temperature lowers the viscosity of the resin, allowing it to flow. Together with low pressure, the volatiles trapped within the laminate are forced out, thus reducing voids. If the viscosity of the resin is too high it will not flow, trapping the volatiles within the laminate, or if too low, then the resin will flow too much, creating areas of resin starvation. It is important that the temperature across the face of the mould tool remains relatively uniform to maintain uniform viscosity of the resin. Therefore, Prudham [18] performed thermal imaging of the mould tool to gain insight into the temperature uniformity across the mould tool surface. Figure 9 shows the thermal imaging of the mould tool surface performed by Prudham [18].

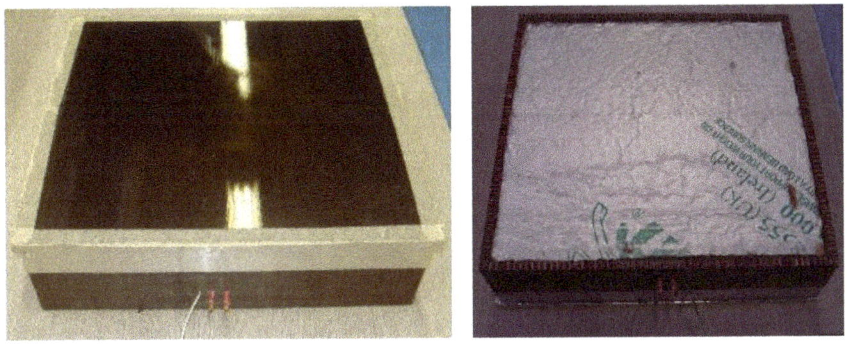

Figure 8. Heated mould tool used for cool-clave processing (acquired with permission from [18]).

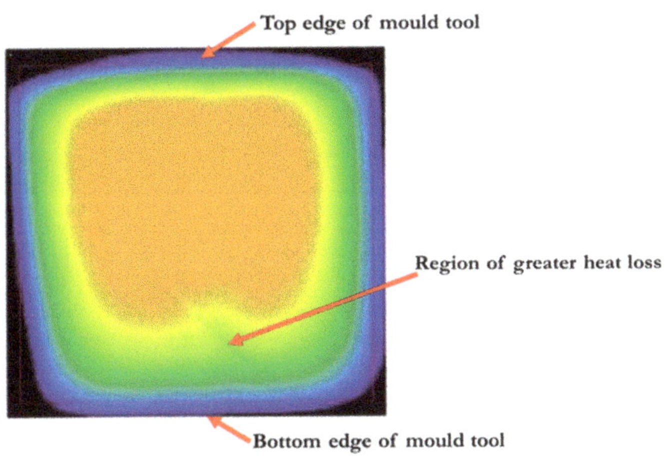

Figure 9. Thermal imaging of mould tool surface (acquired with permission from [18]).

The results indicated discrepancy between temperatures recorded by the mould tools built-in thermocouple and the temperatures recorded during thermal imaging technique. A greater heat loss near to the edges of the mould tool was recorded by thermal imaging with a maximum temperature difference across the main region of the mould tool being in the range from 5 °C horizontally and 7 °C vertically [18] in the image.

From the thermal imaging results, it was concluded that the mould tool would be suitable for cool-clave processing, demonstrating a convective heat transfer across the surface [18]. Greater heat loss across the edges of the mould tool surface was explained by in-plane anisotropic thermal conductivity of carbon fibre-reinforced epoxy composite used in the study [18,28]. The tests highlighted the need for adequate insulation of the mould tool during the manufacturing stage to prevent greater heat losses as recorded during thermal imaging [18].

A Thermal Insulation Hud (TIH) was constructed to prevent any convective heat loss from the mould tool which would encase the mould tool, allowing vacuum and thermocouple connections but limiting convective heat transfer [18]. One of the main challenges of the cool-clave technique was how to get power to the heated mould tool inside the autoclave. The electrical supply used to power the heated mould tool was provided by a power transformer controlled by an 'IMO' PID controller by using a preexisting access hole on the autoclave chamber, plugged together with a specially made fitting encompassing the electrical wires providing the pressure tight seal required [18]. A schematic of the power supply mechanism for the heated mould tool inside the autoclave is shown in Figure 10.

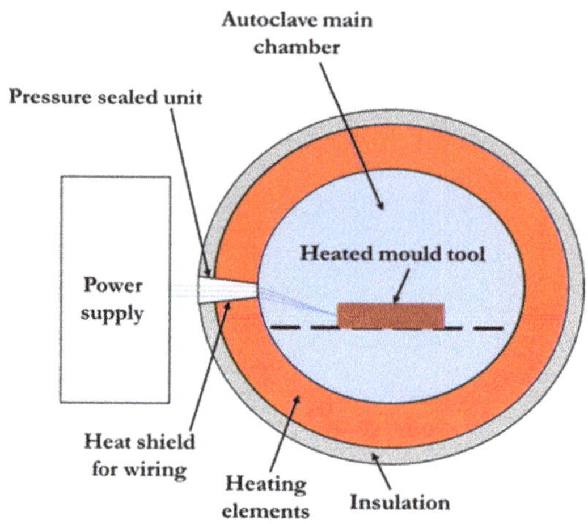

Figure 10. Schematic of the power-supply mechanism for the heated mould tool inside the autoclave chamber (redrawn from [18]).

3.1. Analysis of Energy Consumption between a Traditional Autoclave and Cool-Clave Technique

The main sources of energy consumption in a traditional autoclave and cool-clave process are compressor, and main controls (for both processes), heating elements in traditional autoclave, and heated mould tool in the cool-clave process. The principal objective behind cool-clave process is to achieve autoclave consolidation of composite laminates but being more energy- and cost-efficient. Prudham and Summerscales [27] quantified the energy consumption in a cool-clave processing technique during a defined cure cycle and compared it with a traditional autoclave process. The energy consumption by control units and compressor was deemed to be approximately equal for the traditional autoclave processing and cool-clave technique, as both processes require the autoclave to run with a pressurised chamber. To make a clear distinction between the energy consumption of both processes, the amount of energy required to provide heat during a defined curing cycle was evaluated [27]. The results demonstrated a 35% reduction in energy requirement for heating the laminate when replacing the vessel heaters with heated tooling [27].

The cure cycle in a trditional autoclave requires the component to enter the vessel at time 0 to start the dwell stage. Figure 11a illustrates a traditional autoclave cure cycle. Following thermodynamic analysis based on a traditional autoclave cure cycle, the total energy consumption reported by Prudham and Summerscales [27] was 3620 kJ. Table 1 further illustrates the results demonstrated in the study [27]. The cycle that an autoclave goes through to cure a laminate was split into six phases for the process of analysis. The analysis was carried out assuming the cure cycle of a 12.7 mm thick Cytec Cycom 5216 epoxy prepreg [27].

1. heat from room temperature (20 °C) up to 80 °C
2. pressurisation of autoclave at 80 °C to 0.7 MPa (gauge)
3. reheat to 80 °C to compensate for the temperature drop during pressurisation
4. hold temperature for 30 min
5. heat from 80 °C up to 130 °C
6. hold temperature for 95 min

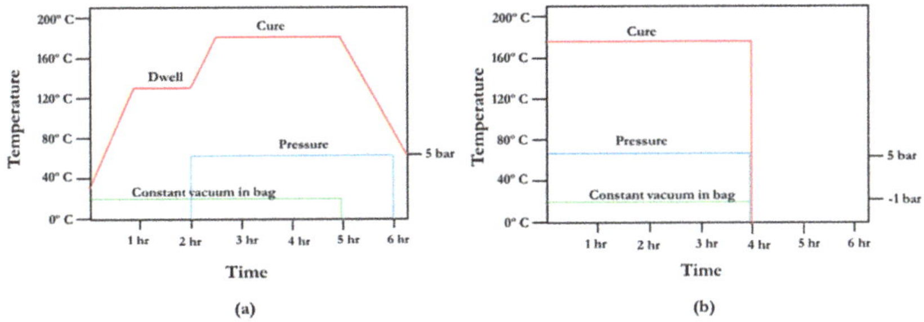

Figure 11. (**a**) Traditional autoclave cure cycle, (**b**) cure cycle for cool-clave processing (redrawn from [29]).

Table 1. Estimated energy consumption of a traditional autoclave cure cycle (reproduced from [12,27]).

Phase	Start Temperature (°C)	Final Temperature (°C)	Energy Consumed (kJ)
1	20	80	157.5
2	80	27.7	0
3	27.7	80	1061
4	80	80	306.4
5	80	130	1014.2
6	130	130	1081.2
	Total energy consumed		3620.3

The total volume of pressurised gas inside the autoclave chamber was estimated to be 1 cubic metre through thermodynamic analysis [27], which was comprised of the cylindrical working space of the autoclave, annular heating channels, and domed ends.

Figure 11b illustrates the cure cycle for cool-clave processing technique [29]. In the proposed cool-clave cure cycle, the component can complete the dwell period and reach full heat prior to entering the vessel. This reduces the time required to be inside the vessel, thus increasing the production rate of components. Additionally, quality is key for acquiring the highest possible safety while minimising costs. Composites are excellent for tailoring for the loads and areas with stress concentrations and can provide tailorable tensile strength [29,30].

With the proposed method, the cycle times in industrial production line can be significantly reduced, which can bring a number of benefits, such as [29]:

1. load the autoclave after the dwell period
2. higher productivity
3. optimise profit potential
4. cure multiple components with different temperatures and cycle times
5. increase in the potential of new innovative products
6. less energy required for the process, a possibility of the use of clean energy

Following cool-clave processing using the heated mould tool technique, the maximum energy consumption based on processing of three Cytec Cycom 5216 560 gsm non-crimp glass fibre-reinforced epoxy composites with ~50% fibre volume fractions was estimated to be 2340 kJ, demonstrating a 35% reduction in energy consumption [12,27].

3.2. Consolidating Resin-Infused Laminates inside Autoclave

Lewin [31] conducted initial experiments as a brief feasibility study to investigate the possibility of consolidating resin-infused laminates inside an autoclave following RIFT II (Resin Infusion under Flexible Tooling) manufacturing outside the autoclave. Laminate (Plate E) properties were compared with laminates manufactured using hand lamination with edge dams to constrain the flow of the infusion resin (Plate A), resin infusion under flexible tooling with a flow medium (RIFT II) and 0.03 MPa (Plate B), 0.06 MPa (Plate C), or 0.09 MPa (Plate D) net pressure [12,31]. Each laminate was manufactured using 270 gsm plain woven glass fabric-infused with IP2 polyester infusion resin (initial viscosity 1600 mPa.s at 25 °C, according to the manufacturer's datasheet) and 2% Butanox M50 MEKP catalyst by weight [12]. Laminate characterisation was performed by undertaking (a) resin burn-off for fibre volume fraction (V_f), (b) tensile properties (BS EN ISO 527-4) (c) flexural properties (BS EN ISO 14125 Class III), (d) inter-laminar shear strength (ILSS BS EN ISO 14130), and (e) surface-breaking voids (SBV) by filling voids with carbon dust followed by image processing and analysis with ImageJ software. The results are presented in Table 2.

Table 2. Results from each laminate characterization tests [31].

Property	Units	Laminates				
		A	B	C	D	E
Process		Hand lay-up	RIFT II	RIFT II	RIFT II	Autoclave
Net pressure	MPa	0	0.03	0.06	0.09	0.586
Plate thickness	mm	2.34 ± 0.06	2.09 ± 0.04	2.03 ± 0.05	1.94 ± 0.02	1.93 ± 0.03
V_f (thickness)	%	40.8	45.7	47.0	49.3	49.6
V_f (burn-off)	%	42.0	46.5	46.7	48.1	50.9
Young's modulus	GPa	21.9 ± 0.6	23.7 ± 0.3	24.5 ± 0.6	24.9 ± 0.4	25.0 ± 0.2
Flex. modulus 40 mm span	GPa	18.7 ± 0.4	19.8 ± 0.5	19.8 ± 0.5	21.1 ± 0.5	22.4 ± 0.4
Flex. modulus 48 mm span	GPa	17.3 ± 0.2	19.0 ± 0.3	18.8 ± 1.0	20.4 ± 0.3	20.9 ± 0.7
Tensile strength	MPa	384 ± 15	416 ± 20	426 ± 26	452 ± 31	478 ± 24
Flexural strength	MPa	558 ± 7	578 ± 13	586 ± 11	599 ± 21	608 ± 13
ILSS 10.0 mm span	MPa	56.4 ± 0.8	53.5 ± 1.3	54.1 ± 1.4	53.4 ± 1.6	52.5 ± 1.1
ILSS 11.4 mm span	MPa	52.5 ± 1.7	49.9 ± 1.7	51.2 ± 0.6	46.6 ± 0.9	48.7 ± 1.0
SBV area	%	1.9	2.4	1.4	0.3	0.02

The fibre volume fraction, elastic moduli, and tensile and flexural strengths all increased with increasing net pressure during manufacturing. The ILSS decreased with increasing fibre volume fraction. The minimal gain in fibre volume fraction for Panel E was attributed to insufficient volume for resin bleed, with the possibility of a greater increase for an optimised process [12,31].

Stringer [32] identified 7500–16,500 mPa.s as an optimum processing window for application of vacuum for void-free high fibre volume fraction composites manufactured by wet lamination and vacuum bagging techniques. Therefore, with some adaptation of the autoclave pressure vessel, it might be practical to load the vacuum-bagged dry composite into the autoclave, then infuse and cure in situ. Preparation outside the autoclave would permit shorter autoclave cure cycles and better utilisation of the pressure vessel [12].

Improved Autoclave Process for Resin-Infused Laminates

Experiments conducted by Wilkinson [33] with a dry fabric reservoir inside the bag and with both the inlet and outlet pipes clamped gave no significant change in fibre volume fraction (demonstrating only 0.26% increase) when compared to plates cured at ambient pressure. Subsequent tests used no reservoir material with the resin inlet clamped, while the resin outlet was vented to atmosphere during autoclave consolidation, demonstrating an 8.6% increase in fibre volume fraction [12,31,33].

After infusion, plates were subjected to (a) vacuum-bag only pressure, (b) 0.31 MPa pressure in the autoclave, or (c) 0.59 MPa pressure in the autoclave. Further laminates were prepared and pressurised after a dwell period to study the effect of viscosity at the time, and pressure was applied for the four times identified by the viscosity tests. The results are illustrated in Table 3.

Table 3. Summary of data obtained from laminates manufactured by outlet pipe vented to air [33].

Property	Units	Infusion	3.1/0.0	5.9/0.0	5.9/39.5	5.9/45	5.9/48.3
Ext. pressure	MPa	0.0	3.1×10^{-4}	5.9×10^{-4}	5.9×10^{-4}	5.9×10^{-4}	5.9×10^{-4}
Dwell time	min	0.0	0.0	0.0	39.5	45	48.3
Plate thickness	μm	2040 ± 51	1890 ± 3	1780 ± 16	1880 ± 1	1930 ± 11	1980 ± 1
V_f (thickness)	%	51.9 ± 1	56.1 ± 0.1	59.3 ± 0.5	56.3 ± 0.0	54.7 ± 0.3	53.6 ± 0.0
V_f (burn-off)	%	52.3	56.4	60.9	56.7	54.3	52.4
Flexural modulus	GPa	20.5 ± 0.4	25.6 ± 0.9	28.4 ± 0.9	26.4 ± 0.7	25.4 ± 0.5	24.0 ± 1.0
Flexural strength	MPa	347 ± 14	384 ± 16	415 ± 20	391 ± 14	383 ± 10	364 ± 12
ILSS	MPa	41.0 ± 1.9	44.5 ± 2.5	41.5 ± 2.5	42.5 ± 3.4	43.2 ± 1.8	42.4 ± 1.7

At constant consolidation pressure, delaying the consolidation resulted in a lower fibre volume fraction in the composite panel. Flexural strength increased with increasing consolidation pressure. Increased viscosity limited the quantity of resin expelled from the laminate, reduced the fibre volume fraction, and resulted in lower mechanical properties [33].

4. Conclusions

The out-of-autoclave manufacturing technique has limitations over maximum achievable fibre volume fraction in composites due to compressibility characteristics of the reinforcement material. Lower fibre volume fraction inevitably results in increased matrix volume fraction and consequent resin-rich volumes. Fibre clustering and RRV cause reductions in composite strength. Autoclave processing of composites is required to achieve the highest performance composites systems for industrial applications in the aerospace, automotive, and defence sectors.

Energy savings can result from decoupling the heat and pressure during the autoclave processing of composites. The use of electrically heated mould tools could eliminate heating of the parasitic systems (pressure vessel walls, insulation, and heat transfer gasses). The consolidation pressure can then be supplied using cool air.

The adoption of heated tooling in the cool-clave technique to achieve autoclave consolidation of composites can significantly reduce process cycle times, as the composites can be taken to the end of dwell period before loading the pressure vessel.

The use of resin infusion, rather than expensive pre-impregnated reinforcements, removes the need for the separate impregnation stage and elimination of power requirements for freezer storage of prepreg materials. Autoclave loading efficiency could be improved by curing different composite systems simultaneously with the composites brought to their respective curing temperatures before loading the autoclave, which would further enhance process efficiency.

Author Contributions: Conceptualization, J.S.; writing—original draft preparation, I.R.C., writing—review and editing, I.R.C. and J.S. All authors have read and agreed to the published version of the manuscript.

Funding: This research received no external funding.

Institutional Review Board Statement: Not applicable.

Informed Consent Statement: Not applicable.

Data Availability Statement: No new data was created.

Conflicts of Interest: The authors declare no conflict of interest.

References

1. Mahmood, A.S.; Summerscales, J.; James, M.N. Resin-rich volumes (RRV) and the performance of fibre-reinforced composites: A review. *J. Compos. Sci.* **2022**, *6*, 53. [CrossRef]
2. Summerscales, J. Fibre Distribution and the Process-Property Dilemma. In *The Structural Integrity of Carbon Fibre Composites: Fifty Years of Progress and Achievement*; Beaumont, P.W.R., Soutis, C., Eds.; Springer: Berlin/Heidelberg, Germany, 2016; pp. 301–317.
3. Centea, T.; Grunenfelder, L.K.; Nutt, S.R. A review of out-of-autoclave prepregs—Material properties, process phenomena, and manufacturing considerations. *Compos. A* **2015**, *70*, 132–154. [CrossRef]
4. Okunzuwa, A.E.; Anjum, N.; Eze, V.O.; Okoli, O.I. A Review on the Out-of-Autoclave Process for Composite Manufacturing *J. Compos. Sci.* **2022**, *6*, 172–202.
5. Solan, J. Out-of-Autoclave Processing for Aero Use, Composites World Supplement, 2016. Available online: https://www.compositesworld.com/articles/february-supplement-out-of-autoclave-processing-for-aero-applications (accessed on 29 December 2022).
6. Arney, M.; Grove, S.M.; Progoulakis, I.; Searle, T.; Short, D.; Spooner, J.; Summerscales, J. Integrally Heated Tooling for the Manufacture of Fibre-Reinforced Composites. In Proceedings of the Conference 'Composites Processing 2004', CPA, Bromsgrove, UK, 28–30 April 2004.
7. Williams, C.D.; Summerscales, J.; Grove, S.M. Resin Infusion under Flexible Tooling (RIFT): A review. *Compos. A* **1996**, *27*, 517–524. [CrossRef]
8. Summerscales, J.; Searle, T.J. Low pressure (vacuum infusion) techniques for moulding large composite structures. *Proc IMechE Part L J. Mater. Des. Appl.* **2005**, *219*, 45–58. [CrossRef]
9. Summerscales, J. Resin Infusion Under Flexible Tooling (RIFT). In *Encyclopedia of Composites*, 2nd ed.; Nicolais, L., Borzacchiello, A., Eds.; John Wiley & Sons: Hoboken, NJ, USA, 2012; pp. 2648–2658.
10. Fox, B.L.; Herring, M.L. The effect of a rapid curing process on the surface finish of a carbon fibre epoxy composite. *Compos. B* **2011**, *42*, 1035–1043.
11. Bogetti, T.A.; Gillespie, J.W. Two-Dimensional Cure Simulation of Thick Thermosetting Composites. *J. Compos. Mater.* **1991**, *25*, 239–273. [CrossRef]
12. Summerscales, J. Taking the high energy demand out of autoclave processing of composites. In Proceedings of the SAMPE European Conference, Southampton, UK, 11–13 September 2018.
13. Dayananda, G.N.; Subba Rao, M.; Somashekar, B.R. Indigenous development of autoclave technology. In Proceedings of the National Symposium on Development in Advanced Composites and Structures, Hyderabad, India, 17 September 1994.
14. Advani, S.G.; Sozer, E.M. Processing advanced thermoset fiber composites. In *Process Modelling in Composites Manufacturing*; Marcel Dekker: New York, NY, USA, 2002; pp. 339–343.
15. Monaghan, M.R.; Mallon, P.J. Development of a computer-controlled autoclave for forming thermoplastic composites. *Compos. Manuf.* **1990**, *1*, 8–14. [CrossRef]
16. Stratton, P.F.; Groome, D.G. BOC gases, enhanced safety in advanced polymer composite curing autoclaves. In Proceedings of the 42nd International SAMPE Symposium and Exhibition, Anaheim, CA, USA, 4–8 May 1997; pp. 172–180.
17. Upadhya, A.R.; Dayananda, G.N.; Kamalakannan, G.M.; Setty, J.R.; Daniel, J.C. Autoclaves for Aerospace Applications: Issues and Challenges. *Int. J. Aerosp. Eng.* **2011**, *2011*, 1–11. [CrossRef]
18. Prudham, G. Development of Cool Clave Processing Method. Bachelor's Thesis, University of Plymouth, Plymouth, UK, 2010.
19. Kamalakannan, G.M.; Rao, M.S. Development of a computer-based process control system for an autoclave to cure polymer matrix composites. In Proceedings of the International Conference on Instrumentation, Pune, India, 19–21 December 2004.
20. Alam, P. *Composites Engineering: An A-Z Guide*; IOP Publishing: Bristol, UK, 2021.
21. Eckold, G. *Design and Manufacture of Composite Structures*; Woodhead Publishing: Cambridge, UK, 1994.
22. Fernandez, I.; Blas, F.; Frovel, M. Autoclave forming of thermoplastic composite parts. *J. Mater. Process. Technol.* **2003**, *143–144*, 266–290. [CrossRef]
23. Dhakal, H.M.; Ismail, S.O. Design, manufacturing processes and their effects on bio-composite properties. In *Sustainable Composites for Lightweight Applications*; Dhakal, H.M., Ismail, S.O., Eds.; Series in Composites Science and Engineering; Woodhead Publishing: Sawston, UK, 2021; pp. 121–177.

24. Gupta, M.; Jain, A.; Kamineni, J.N.; Burela, R.G. Advances and applications of biofiber-based polymer composites. In *Advances in Bio-Based Fiber*; Rangappa, S.M., Puttegowda, M., Parameswaranpillai, J., Siengchin, S., Gorbatyuk, S., Eds.; Woodhead Publishing: Sawston, UK, 2022; pp. 575–602.
25. Kamalakannan, G.M.; Gupta, A.K. An improved technique and its implementation for control of high-power heaters in large autoclaves and similar plants. In Proceedings of the International Conference of Instrumentation, Pune, India, 17–19 December 2009.
26. Kamalakannan, G.M. Ethernet based online process monitoring and controlling of a plant having RS-232 or RS-485 equipment. In Proceedings of the National Symposium on Instrumentation, NSI-32, Erode, India, 24–26 October 2007.
27. Prudham, G.; Summerscales, J. Cost reduction in autoclave processing I: Heated tooling and cool air pressurisation. In Proceedings of the 13th International Conference on Flow Processing in Composite Materials, Tokyo, Japan, 6–9 July 2016.
28. Grove, S.M.; Progoulakis, I.; Searle, T.J.; Summerscales, J.; Healey, P. Heated Tooling for Aerospace Composites Manufacture. *SAMPE J.* **2005**, *41*, 36–45.
29. Snow, J.C. Thermal Analysis of Composites Manufacture in the Autoclave with Heated Tooling and Cool Air Pressurisation. Bachelor's Thesis, University of Plymouth, Plymouth, UK, 2018.
30. Ciriscioli, P.R.; Springer, G.S. *Smart Autoclave Cure of Composites*; Technomic Publishing Company: Lancaster, PA, USA, 1990.
31. Lewin, G. The Influence of Resin Rich Volumes (RRV) on Composite Mechanical Properties. Bachelor's Thesis, University of Plymouth, Plymouth, UK, 2016.
32. Stringer, L.G. Optimization of the wet lay-up/vacuum bag process for the fabrication of carbon fibre epoxy composites with high fibre fraction and low void content. *Composites* **1989**, *20*, 441–452. [CrossRef]
33. Wilkinson, R. Enhancing the Mechanical Properties of A Resin Infused Composite Using Autoclave Consolidation. Bachelor's Thesis, University of Plymouth, Plymouth, UK, 2017.

Journal of
Composites Science

MDPI

Article

On the Machining Temperature and Hole Quality of CFRP Laminates When Using Diamond-Coated Special Drills

Jinyang Xu [1,*], Tieyu Lin [1] and Joao Paulo Davim [2]

[1] State Key Laboratory of Mechanical System and Vibration, School of Mechanical Engineering, Shanghai Jiao Tong University, Shanghai 200240, China; lintieyu@sjtu.edu.cn
[2] Department of Mechanical Engineering, University of Aveiro, Campus Santiago, 3810-193 Aveiro, Portugal; pdavim@ua.pt
* Correspondence: xujinyang@sjtu.edu.cn

Abstract: Carbon fiber reinforced polymers (CFRPs) are attractive engineering materials in the modern aerospace industry, but possess extremely poor machinability because of their inherent anisotropy and heterogeneity. Although substantial research work has been conducted to understand the drilling behavior of CFRPs, some critical aspects related to the machining temperature development and its correlations with the process parameters still need to be addressed. The present paper aims to characterize the temperature variation and evolution during the CFRP drilling using diamond-coated candlestick and step tools. Progression of the composite drilling temperatures was recorded using an infrared thermography camera, and the hole quality was assessed in terms of surface morphologies and hole diameters. The results indicate that the maximum drilling temperature tends to be reached when the drill edges are fully engaged into the composite workpiece. Then it drops sharply as the tool tends to exit the last fiber plies. Lower cutting speeds and lower feed rates are found to favor the reduction of the maximum composite drilling temperature, thus reducing the risk of the matrix glass transition. The candlestick drill promotes lower magnitudes of drilling temperatures, while the step drill yields better surface morphologies and more consistent hole diameters due to the reaming effects of its secondary step edges.

Keywords: CFRP composites; drilling process; special drills; machining temperatures; hole quality

Citation: Xu, J.; Lin, T.; Davim, J.P. On the Machining Temperature and Hole Quality of CFRP Laminates When Using Diamond-Coated Special Drills. *J. Compos. Sci.* **2022**, *6*, 45. https://doi.org/10.3390/jcs6020045

Academic Editor: Francesco Tornabene

Received: 18 January 2022
Accepted: 29 January 2022
Published: 1 February 2022

1. Introduction

In recent decades, carbon fiber reinforced polymers (CFRPs) have been receiving immense attention in diverse engineering fields due to their superior properties and unique functionality [1–4]. This can be seen by their widespread applications for fabricating the main load-bearing components in the modern aerospace industries. For instance, CFRPs are extensively used in the wing boxes, horizontal and vertical stabilizers, and wing panels of large commercial aircrafts, such as Airbus A380 and Boeing 787 [4,5]. Generally, CFRPs feature two typical constituents, namely reinforcing fibers and an impregnating matrix, which show completely disparate behaviors [2,6]. The composites are characterized by high specific mechanical/physical properties, being a promising alternative to conventional metallic alloys and steels. Contrary to isotopic materials, CFRP composites generally exhibit a heterogeneous structure and anisotropic behavior, being regarded as a rather difficult-to-cut material. Although most CFRP components are fabricated to near-net shapes by molding processes, mechanical machining has become a compulsory operation in order to achieve desired dimensional accuracy and target quality attributes for final composite products [7–10]. However, the inherent anisotropy and heterogeneity of fibrous composites complicate the chip separation process and tend to cause extremely undesirable machining consequences such as severe surface damage, rapid tool wear, increased cutting costs. Meanwhile, the machinability of CFRPs is fiber-orientation dependent, owing to the varying fiber fracture mechanisms associated with the fiber cutting angle. It has been reported by previous investigations that interrelated chip separation modes such as bending-induced fractures, shear-induced fractures, fiber buckling, and interfacial debonding occur for the cutting of fibrous composites [11–14]. These unique characteristics result in the poor machinability of CFRP materials, posing tremendous challenges to modern manufacturing sectors. Among the secondary machining operations, drilling is the most frequently-used operation for the cutting of CFRPs due to the need to create boreholes for assembling different composite parts into final products. It is roughly estimated that the range of the number of holes required by a commercial aircraft is up to 1.5–3 million, while a jet fighter requires as many as 300,000 holes [4,6]. Thus, from the manufacturers' point of view, the drilling process becomes essential in the final acceptance of composite parts. However, it is rather challenging to drill CFRPs with desired hole quality and target dimensional accuracy. This is due to the varying chip removal modes associated with the changeable fiber cutting angle and the high abrasiveness of the reinforcing fibers. Some of the critical issues encountered in CFRP drilling include drilling-induced delamination, glass transition failure of the composite matrix, poor dimensional accuracy, rapid tool wear, etc. Apart from the mechanical force effects, cutting heat and resulting temperature development also play a vital role in the surface integrity of CFRP materials. In particular, high temperatures can be easily accessed in the drilling processes due to the semi-enclosed environment of the chip separation, leading to poor heat dissipation. Excessive drilling temperatures promoted at the tool–composite interface can cause severe degradations of the composite properties, debonding of the fiber-matrix interfaces, and the glass transition of the matrix base [15–18]. Therefore, studies dealing with temperature initiation and progression are very meaningful to achieve active control of the thermal effects for the drilling of CFRP materials.

To address the drilling issues of CFRPs, a large amount of research work has been carried out worldwide by scholars [10,19–35], covering a variety of aspects involving drilling forces, drilling-induced damages, tool wear, etc. Some drilling-induced damage, including delamination, hole dimensional inaccuracy, surface roughness, fiber pullouts, and uncut fibers, have been investigated in the scientific literature [19,20]. For instance, Davim and Reis [21,22] were among the earliest to deal with the drilling behavior of CFRP composite laminates. In their work, the correlations between the cutting velocity and feed rate with the machining power, specific cutting pressure, and delamination factor were established. The authors stated that both the cutting speed and the feed rate positively affected the progression of the delamination factor, and the use of brad spur drills favored the reduction of drilling-induced delamination. Bonnet et al. [24] studied the local feed force and its consequences on the exit hole damage during CFRP drilling. It was found that the fiber cutting modes changed dynamically with the composite sequence, and the local feed forces generated on the hole bottom could be correlated with the delaminating aspects. Su et al. [28] addressed the thrust forces and delamination issues when drilling CFRPs using a tapered drill-reamer and found that the drilling parameters significantly affected the maximum thrust force measured in the drilling stage instead of the reaming stage. Ameur et al. [29] conducted drilling studies on CFRP materials using different types of tool materials. The authors stated that both the drilling thrust force and the delamination factor were primarily influenced by the tool material and the feed rate, while the hole cylindricity errors were mainly affected by the spindle speed. Rawat and Attia [31] studied the wear behavior of carbide tools during the high-speed drilling of CFRP laminates, and observed that chipping and abrasion were the main wear modes controlling the deterioration of carbide drills. Faraz et al. [32] highlighted the wear phenomenon of cutting-edge rounding (CER) for CFRP drilling and introduced the CER value for the quantification of drill wear. Wang et al. [33] investigated the wear progression of coated tools while drilling CFRP laminates. The authors pointed out that the dominant wear type was dulling or blunting of the cutting edge during CFRP drilling. The use of a diamond coating could significantly reduce the edge rounding wear, while the AlTiN coating failed to protect the drill due to its oxidation during machining. A critical review conducted by Ismail et al. [34] offered a clear understanding of the current advances in drilling composite materials, which focused on the aspects of tool geometries, materials, and parametric designs. Fu et al. [35] and Kubher et al. [36] investigated the temperature characteristics in drilling unidirectional (UD) and multidirectional (MD) CFRPs. Due to the associated temperature effects, utilizing MD CFRPs could result in more difficulties in achieving high drilling qualities than UD CFRPs at certain fiber cutting angles. Through the literature survey, although substantial research work has been conducted to understand the drilling behavior of CFRP composites most of the studies are focused on the analysis of force-related effects, such as drilling thrust forces, delamination damage, hole quality, and tool wear issues. Even though there are some papers that have already addressed the temperature issues for CFRPs, very limited literature has been reported to deal with the temperature variations in drilling high-strength CFRPs with special drills. Moreover, some critical issues related to the machining temperature development and its correlations with the drilling parameters still need to be carefully addressed. Hence, the current work performed in the paper can supplement expertise and knowledge regarding the drilling temperature issues for high-strength CFRP composites. Its novelty lies in identifying the evolution law of the drilling temperature following the entire composite machining operation and in clarifying the parametric effects on the temperature development. Moreover, a particular focus is placed on the evaluation of different special drills for CFRP drilling and on the quantification of hole geometrical accuracy under varying drilling conditions. The experimental results were discussed with respect to the process parameters used. The paper is intended to offer a better understanding of the thermal behavior of CFRP laminates when subjected to the drilling operation.

2. Experimental Procedures

In the present work, machining studies were conducted on the multidirectional CFRP laminates fabricated by high-strength T700 carbon fibers and FRD-YZR-03 epoxy resin. The main composition and basic mechanical properties of the examined CFRP laminates are summarized in Tables 1 and 2, respectively. The composite plate had a total size of 300 mm (length) × 200 mm (width) × 6.60 mm (thickness), which was fabricated by the hand lay-up molding technology. The drilling experiments were performed on a DMU 70 V CNC machining center following a full factorial design of experiments by using candlestick and step drills. The experimental setup for the drilling tests is shown in Figure 1. The input process parameters consist of three levels for the cutting speed (V_c = 40, 80, and 120 m/min) and three levels for the feed rate (f = 0.06, 0.09, and 0.12 mm/rev).

Table 1. The composition of the used CFRP composite.

Reinforcement	Matrix Base	Fiber Volume Fraction	Fiber Bundles
T700 carbon fibers	FRD-YZR-03 epoxy	60%	7 μm, 12 K

Table 2. The mechanical properties of the used CFRP composite.

Tensile Modulus	Tensile Strength	Poisson's Ratio, ν	Flexural Modulus	Flexural Strength	Shear Strength	Glass Transition Temperature
240 GPa	4900 MPa	0.30	210 GPa	1500 MPa	125 MPa	125~135 °C

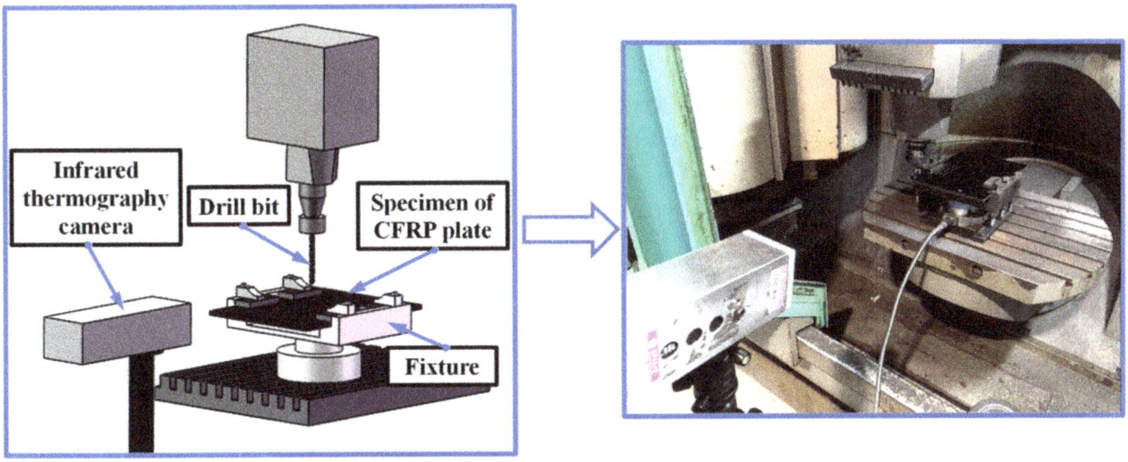

Figure 1. The experimental setup for the CFRP drilling.

Both drills were diamond-coated special tools featuring an 8.0 mm diameter, a 30° helix angle, and a 90° point angle dedicated to delamination suppression and anti-abrasion wear during composite drilling. The detailed morphologies of the candlestick and step drills are shown in Figure 2. The candlestick drill featured three protruding tips, including one centering tip and two peripheral tips, which could significantly reduce the drilling thrust force and ensure the sharp flank cutting edges, whereas the step drill was designed following a step-control scheme involving a first step to create a pilot hole and a secondary step to ream the hole surface to the final diameter. Moreover, the examined step drill featured a ratio of the primary diameter (7.8 mm) to the second diameter (8.0 mm) of 0.975. The small difference between the primary and secondary diameters mainly aimed to let the secondary step edges have a reaming action on the previously cut hole surfaces by the first step drill, as a very small chip removal volume is involved in such step drilling. During the drilling operation, the FLIR A615 infrared thermography camera (IFTC), which featured a working temperature from −20 to 2000 °C and an image acquisition frequency from 50 to 200 Hz, was utilized to in-situ record the temperature development under varying drilling conditions. A similar method was applied by Xu et al. [37], which proved that the temperature measurement chain in the current experiment is capable of measuring the changes in the cutting temperature. The temperature resolution of the equipment was less than 0.05 °C, which guaranteed the accuracy and reliability of the monitored data. The accurate measurements of the cutting temperature by the thermographic camera were also carefully guaranteed by the calibration of emissivity value parameters set in the software. Additionally, an emissivity value of 0.85 was adopted for the composites drilling, according to the recommendations of the infrared thermography camera manufacturer. Moreover, to make the temperature measurements more reliable, all of the composite holes were drilled with a 1.0 mm distance close to the edge of the workpiece. After the completion of the drilling operations, the hole wall morphologies were characterized using the ZEISS confocal laser scanning microscope (CLSM). Finally, the average diameters at the entrance, middle, and exit sides of CFRP holes were measured using a SOLEX EUA coordinate measuring machine (CMM). The obtained results were correlated with the drill bits and the input process parameters.

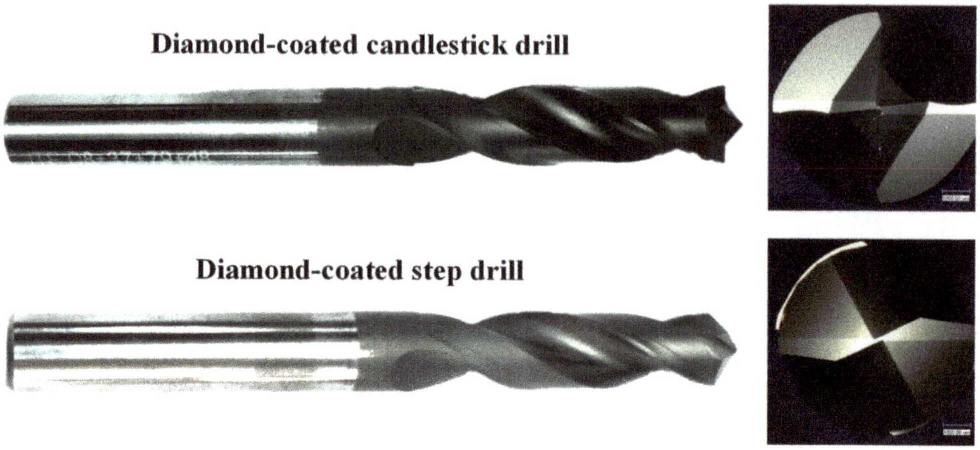

Figure 2. The morphologies of the used special drills.

3. Results and Discussion

3.1. Characterization of Drilling Temperatures

Machining temperature is a characteristic phenomenon of heat accumulation resulting from the tool–work interaction following the material separation process. When dealing with the hole-making processes of CFRP composites, drilling temperature is a critical issue that has to be carefully addressed, as high levels of temperatures can be easily accessed due to the poor heat dissipation of the drill–composite interaction. Additionally, high temperatures are extremely detrimental to the surface integrity and mechanical properties of cut composite holes as they can cause degradation of the composite properties, debonding of the fiber/matrix interface, or even the glass transition failure of the matrix base. Therefore, it is essential to characterize the variation laws of the temperature rise and progression during the machining of CFRP laminates. Figures 3 and 4 show the recorded thermal images of the drilling temperature development under varying feed rates for the candlestick and step drills, respectively, during CFRP machining. It is clear that the temperature progression characteristics can be divided into three stages with respect to the tool–work interaction At the early stage, the drill edges start to attack the composite laminate, and a large amount of cutting heat is progressively generated through the tool–chip and tool–work interactions. As brittle fracture dominates the chip separation of the carbon/epoxy composites, more cutting heat is likely to accumulate within a very narrow tool–chip interface, resulting in a high temperature rise as the drill tends to penetrate inside the composite. At the drill entrance stage, moderate levels of drilling temperatures are produced for both the candlestick and step drills. Meanwhile, with the ongoing tool advancement, the drill edges are fully engaged in the cutting of the fiber/epoxy material. In such circumstances, peak values of drilling temperatures are identified through the thermal image examinations for both drills used. Due to the heat accumulation effects and the full drill interaction with the composite material, the maximum temperature is reached, which indicates the highest risk of occurrence of the thermally-induced damage onto the internal composite hole walls. In most cases, the candlestick drills are found to produce relatively lower values of drilling temperatures than the step drills, particularly under the full drill–work interaction stage, as depicted in Figures 3 and 4. Additionally, increasing the feed rate tends to elevate the maximum temperatures at the full tool engagement stage, due to the increased amount of frictional heat generated as the feed rate rises. Eventually, when the tool retracts from the composite workpiece, the drilling temperature appears to decrease dramatically because of the effective heat dissipation and air cooling of the tool–work system. It is also worth noting that both the candlestick and step drills yield comparable values of drilling temperatures at the drill retraction stage while machining the CFRP laminates.

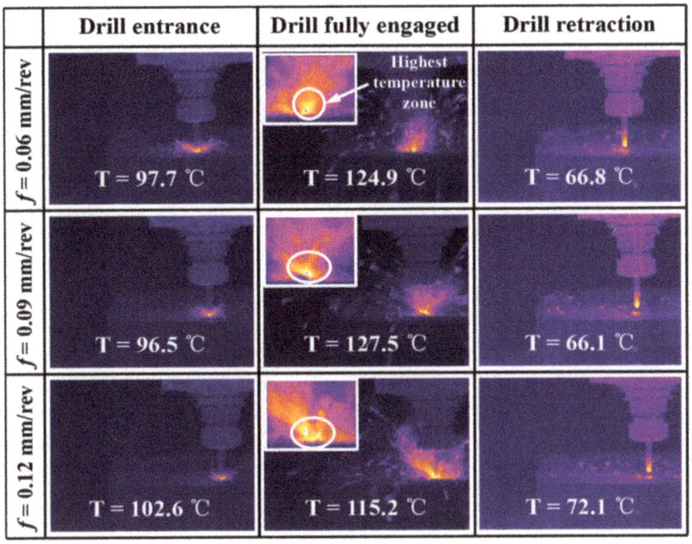

Figure 3. Thermal images of drilling temperature development when using candlestick drills (V_c = 120 m/min).

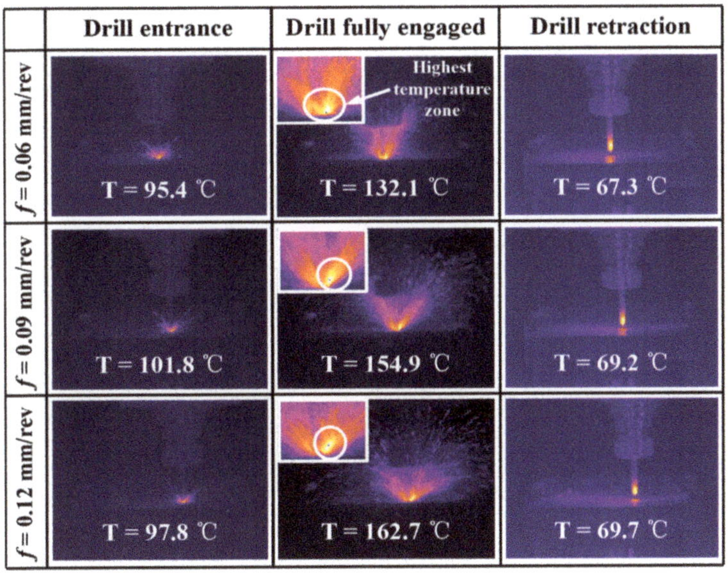

Figure 4. Thermal images of drilling temperature development when using step drills (V_c = 120 m/min).

Figure 5 also shows the comparative evolution of the drilling temperatures in terms of the cutting time (t) following a complete CFRP drilling process for both types of special tools. It is clear that the drilling temperature signals fluctuate significantly during the composite removal process, which exhibits a rapidly increasing trend at the drill entrance stage when the tool starts to attack the composite specimen. Then, the temperature signals for both drills appear to decrease gradually with the tool advancement as the drill edges start to exit the last composite plies, resulting in reduced frictional heat generation following the chip removal process. The previous investigation done by Fu et al. [35] revealed the complex drill-exit temperature characteristics for UD and MD CFRPs, and similar temperature changes were identified in their work. The maximum drilling temperatures for both drills seemed to be reached at around t = 0.5 s under the tested process parameters. The corresponding feed depth of the time is about 2.8 mm after the calculation using the time, feed rate, and cutting speed. Moreover, the candlestick drill is found to promote lower levels of peak drilling temperatures than the step drill. Figure 6 presents the comparison of the maximum temperatures recorded in the CFRP drilling between the two types of drills. Note that the maximum temperatures denoted herein signify the average value of the highest temperature range during the composite drilling. The results also indicate that the candlestick drills generally yield lower drilling temperatures than the step drills for all of the cutting conditions examined. This is due to the two protruding tips of the candlestick drill along the drill periphery that reduce the frictional heat generation and improve the heat dissipation at the tool–chip interactions. Additionally, the cutting speed definitely shows a positive impact on the progression of drilling temperatures for both tools, except for the abnormal temperature data at V_c = 40 m/min for the step drill. The phenomenon is associated with the intensified friction of the tool–composite interaction when the cutting speed increases. Moreover, increasing the feed rate appears to raise the drilling temperatures for the two special drills when machining the CFRP materials.

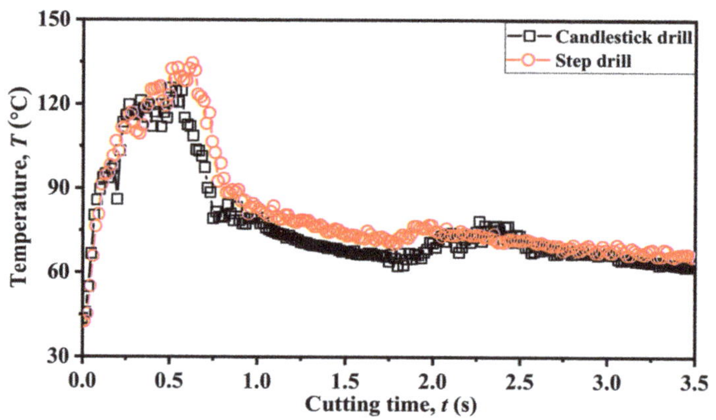

Figure 5. Comparison of drilling temperature development between candlestick and step drills (V_c = 120 m/min and f = 0.06 mm/rev).

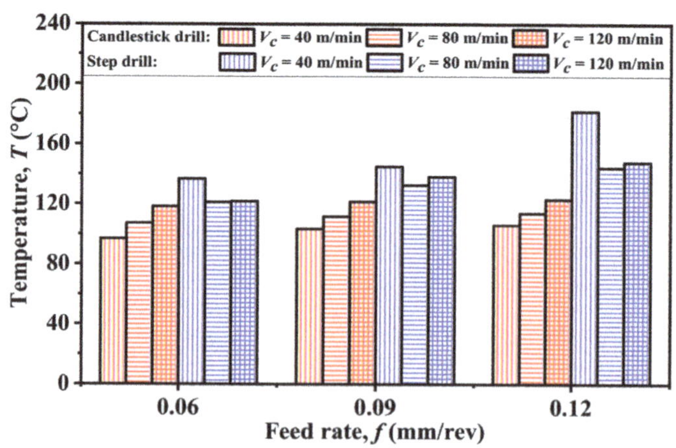

Figure 6. Evolution of the maximum drilling temperatures in terms of the process parameters.

3.2. Hole Wall Morphologies

Machining of fibrous composites differs significantly from the cutting of isotropic metals and steels due to the varying chip separation mechanisms associated with the fiber orientation. The material removal complicates the surface generation of hole walls for the composites under drilling operations. Hole wall morphologies can be considered as one of the most important criteria in assessing the quality attributes of drilled CFRP composites. In general, it is rather difficult to generate smooth surface morphologies, as the material removal mode changes dynamically with the drill rotation during the hole-making process. Figures 7–10 show the topographies of the CFRP hole walls produced by the two types of drills under the fixed cutting conditions (V_c = 120 m/min and f = 0.09 mm/rev). Figures 7–10 all feature the same fiber orientations. From Figure 7, surface flaws due to interlaminar cracking are noted, which feature deep blue colored zones. The finely-cut composite surfaces mainly exist in areas involving the shear-induced fractures of fiber plies. Additionally, the profiles of four circular arc curves at the A–A, B–B, C–C, and L–L cross-sections are plotted in Figures 7 and 9. It is noted that the surface profiles fluctuate significantly along both the radial and axial directions of the holes, which is due to the inherent variations in the surface of the fibers and the matrix. Additionally, the average surface roughness values (R_a) of the selected cross-sections mainly range from 5.00 to 6.29 μm along the hole radial direction, while R_a reaches its maximum value toward the hole axial direction at the L–L cross-section. This is due to the significant disparity in fiber orientation between adjacent fiber plies toward the composite thickness direction.

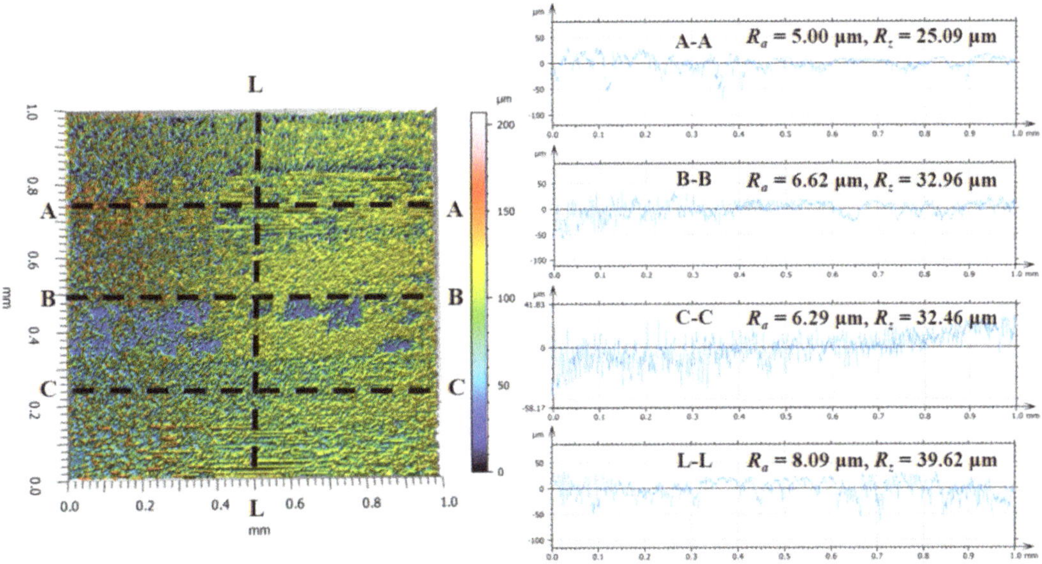

Figure 7. CLSM image of a CFRP hole wall cut by the candlestick drill and its cross-sectional profiles (V_c = 120 m/min and f = 0.09 mm/rev).

With respect to Figure 8, it shows the three-dimensional topographies of cut hole walls at the entrance side. It is evidenced that the cut CFRP hole morphologies feature smooth fiber surfaces containing a certain degree of surface cavities. In contrast, the hole wall morphologies produced by the step drill appear to be much better than those cut by the candlestick drill. As depicted in Figure 9, the R_a values of the four selected cross-sections are relatively lower than those gained by the candlestick drill. The phenomenon is due to the reaming effects of the secondary step edges, indicating the superiority of the step tools in achieving a better hole surface finish than the candlestick tools while machining the CFRP laminates. Moreover, the surface defects residing within the composite hole cut by the step drill mainly include surface cavities due to the loss of matrix, resin smearing, and fiber pullout voids, as shown in Figure 10. Note that the previous research carried out by Kubher et al. [36] addressed the evolution of in-situ cutting temperature and machining forces during the conventional drilling of MD CFRP laminates. Similar surface defects, including resin smearing and bending-induced fracture of carbon fibers, could be found in the research. For both drills in the current investigation, no significant evidence of interlaminar delamination at the hole entrance side is identified through the CLSM examination.

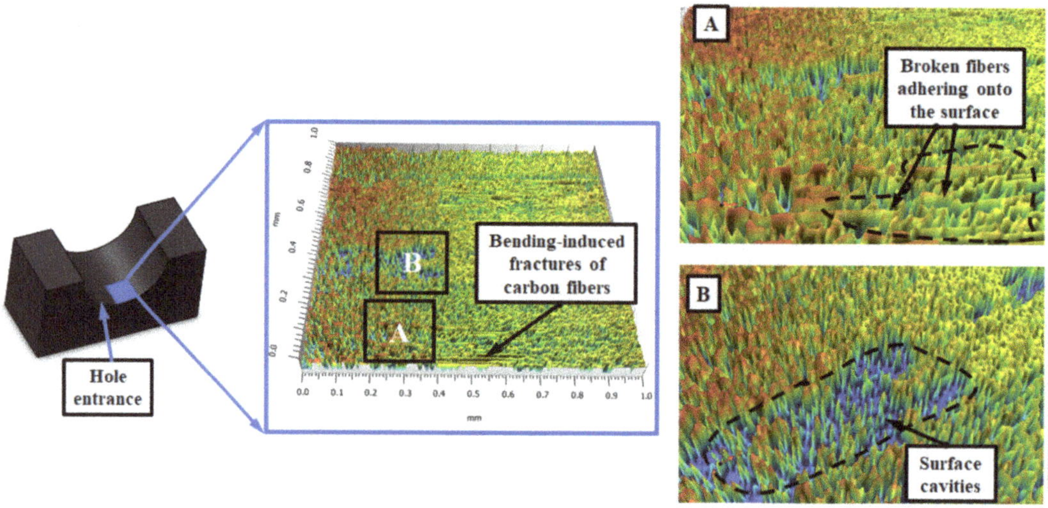

Figure 8. Topographies of a CFRP hole wall cut by the candlestick drill (V_c = 120 m/min and f = 0.09 mm/rev).

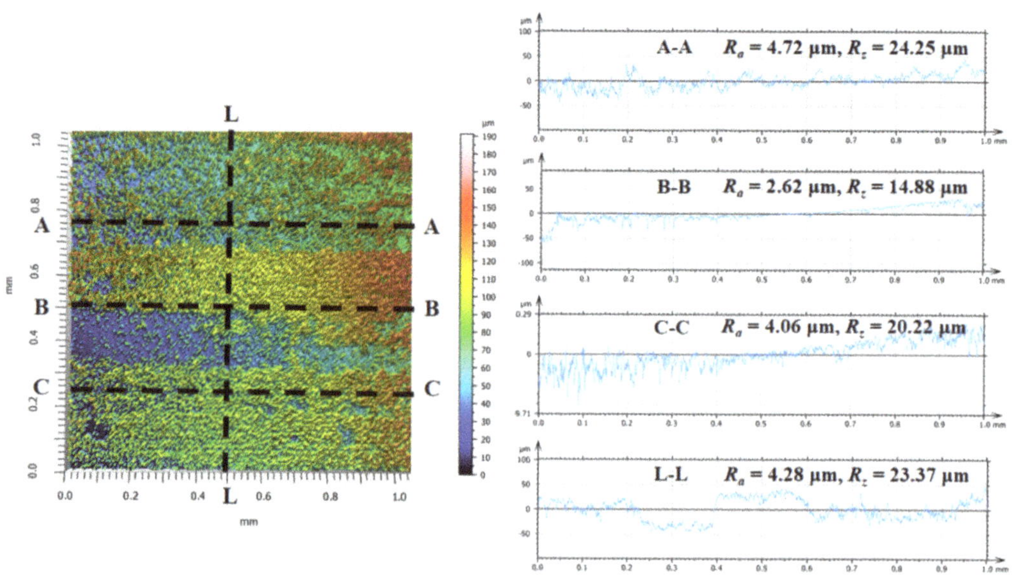

Figure 9. CLSM image of a CFRP hole wall cut by the step drill and its cross-sectional profiles (V_c = 120 m/min and f = 0.09 mm/rev).

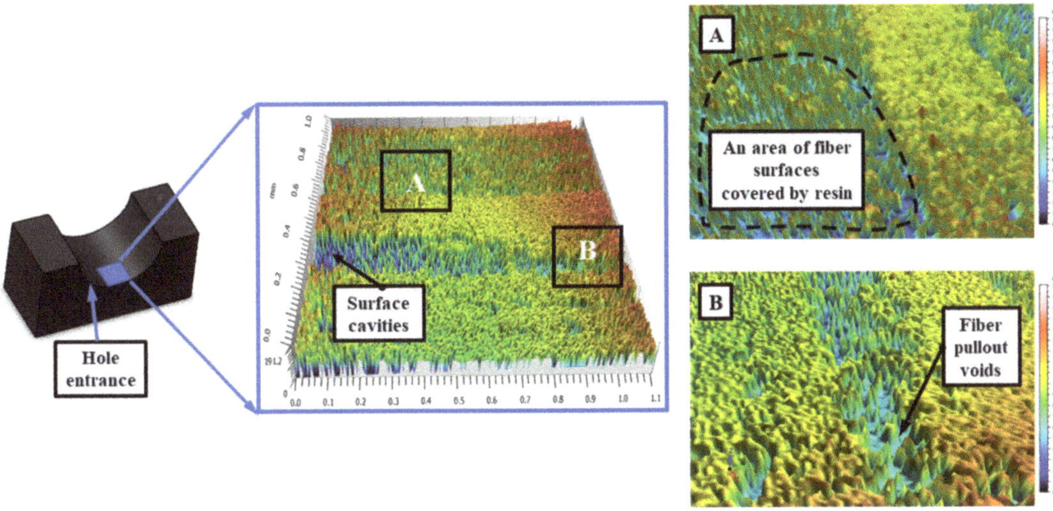

Figure 10. Topographies of a CFRP hole wall cut by the step drill (V_c = 120 m/min and f = 0.09 mm/rev).

3.3. *Hole Diameter*

In drilling CFRP composites, diameter value is an essential criterion for evaluating the hole geometrical accuracy, which determines the assembly performance of the composite parts. In the current work, the average diameters at the entrance, middle, and exit sides of cut CFRP holes were measured and correlated with the process parameters and drill bits used. The obtained results are depicted in Figures 11–13. Both the cutting speed and the feed rate have a significant impact on the variations of the hole diameters, irrespective of the measuring side. Under the lowest speed conditions (V_c = 40 m/min), increasing the feed rate tends to enlarge the cut hole diameters, particularly for the step drills (Figure 11). In most cases, undersized holes are generally produced by the two drills when V_c = 40 m/min. It is worth noting that the diameters measured at the exit side show the largest value, followed by those measured at the middle and entrance sides, regardless of the drill bits and process parameters used. The phenomenon indicates a wedge-shaped cylindrical surface of cut hole walls from the entrance to the exit side due to the intensified tool vibration arising from the decreased stiffness of remaining fiber plies as the fiber layers become much thinner with the tool advancement in drilling. When the moderate speed is used (V_c = 80 m/min), the feed rate fails to show a clear effect on the variations of the hole diameters. In particular, more consistent holes close to the nominal diameter value are promoted by the step drills at the feed rate of 0.09 mm/rev (Figure 12). With respect to the highest speed conditions (V_c = 120 m/min), typically, oversized holes are produced by the candlestick drills, and undersized holes are generated by step drills, as shown in Figure 13. Under such conditions, more consistent holes are created by the candlestick drill at the feed rate of 0.09 mm/rev. Finally, increasing the cutting speed seems to enlarge the hole diameters for the candlestick drills, but tends to decrease the hole diameters for the step drills. In general, to produce consistent holes close to the nominal diameter, the highest speed and moderate feed values are suggested for the candlestick drills, while moderate speed and lower feed values are recommended for the step drills.

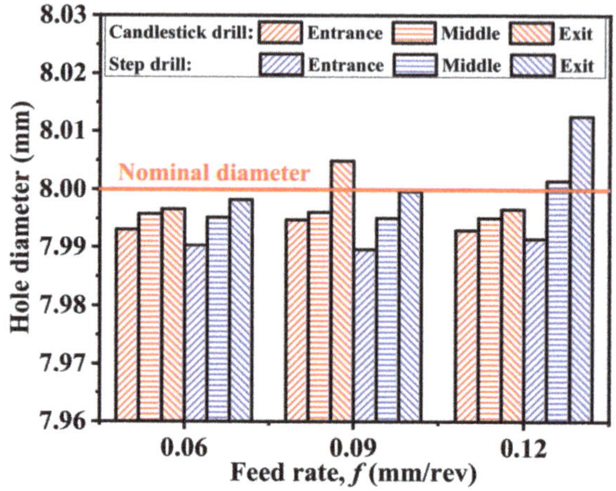

Figure 11. The hole diameter in terms of different feed rates and drill bits (V_c = 40 m/min).

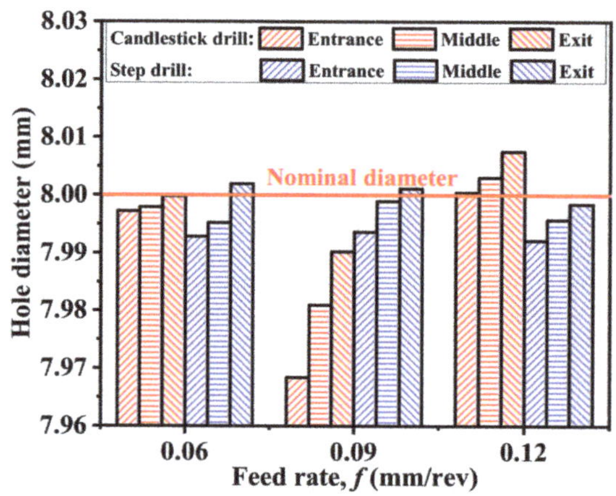

Figure 12. The hole diameter in terms of different feed rates and drill bits (V_c = 80 m/min).

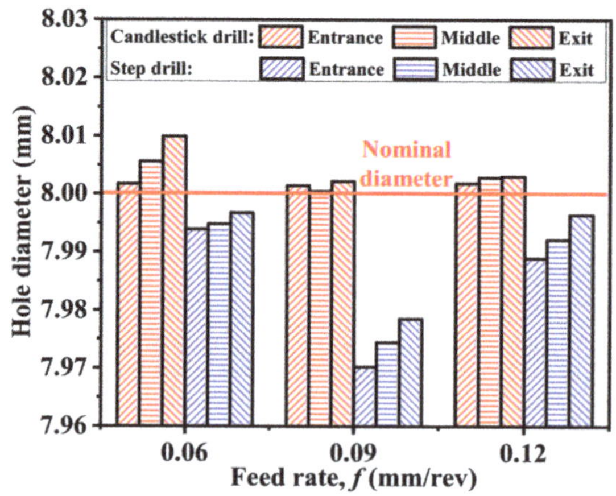

Figure 13. The hole diameter in terms of different feed rates and drill bits (V_c = 120 m/min).

4. Conclusions

This paper deals with the drilling behavior of CFRP composite laminates using diamond-coated special drills. Machining studies were conducted following a full factorial design of experiments. The composite machinability was evaluated, focusing on the machining temperature characteristics and hole quality attributes under varying conditions. The work addresses the temperature variations and progressions during the CFRP drilling. An attempt was made to assess the performances of different special drills and to quantify the hole geometrical accuracy in the CFRP drilling. Based on the results acquired, the following conclusions can be drawn.

- Development of the drilling temperature can be roughly divided into three stages in terms of the tool–work interaction. In general, the maximum temperature values can be attained when the drill edges are fully engaged into the composite workpiece. The progression of the composite drilling temperature basically shows a high sensitivity to the input process parameters. In most cases, both the cutting speed and the feed rate exhibit a positive impact on the temperature rise for all of the drills examined.
- The candlestick drills are found to produce lower magnitudes of drilling temperatures than the step ones. To suppress the temperature progression, low cutting speeds and low feed rates are recommended for the drilling of CFRP composites.
- The cut CFRP hole morphologies are characterized by finely-cut fiber surfaces along with a certain degree of surface cavities due to the loss of matrix. The step drill produces better hole wall morphologies and lower surface roughness values than the candlestick drill due to the reaming effects of its secondary step edges. The main surface defects residing within the cut composite holes include surface cavities, resin smearing, and fiber pullout voids.
- A wedge-shaped cylindrical surface is noted for the cut composite hole wall due to the intensified tool vibration arising from the decreased stiffness of the last fiber plies with the tool advancement. Both the process parameters significantly affect the variations of the hole diameters, irrespective of the measuring side. In general, the highest speed and moderate feed are suggested for the candlestick drills, while the moderate speed and lower feed are recommended for the step drills to create more consistent holes close to the nominal diameter.

Author Contributions: Conceptualization, J.X.; methodology, J.X.; formal analysis, J.X. and T.L.; investigation, J.X.; resources, J.X.; data curation, J.X. and T.L.; writing—original draft preparation, J.X. and T.L.; writing—review and editing, J.P.D.; supervision, J.P.D.; funding acquisition, J.X. All authors have read and agreed to the published version of the manuscript.

Funding: This work was supported by the National Natural Science Foundation of China (grant no. 51705319).

Conflicts of Interest: The authors declare no conflict of interest.

References

1. Abrão, A.M.; Faria, P.E.; Rubio, J.C.C.; Reis, P.; Davim, J.P. Drilling of fiber reinforced plastics: A review. *J. Mater. Process. Technol.* **2007**, *186*, 1–7. [CrossRef]
2. Geier, N.; Davim, J.P.; Szalay, T. Advanced cutting tools and technologies for drilling carbon fibre reinforced polymer (CFRP) composites: A review. *Compos. Pt. A-Appl. Sci. Manuf.* **2019**, *125*, 105552. [CrossRef]
3. Liu, D.; Tang, Y.; Cong, W.L. A review of mechanical drilling for composite laminates. *Compos. Struct.* **2012**, *94*, 1265–1279. [CrossRef]
4. Aamir, M.; Tolouei-Rad, M.; Giasin, K.; Nosrati, A. Recent advances in drilling of carbon fiber–reinforced polymers for aerospace applications: A review. *Int. J. Adv. Manuf. Technol.* **2019**, *105*, 2289–2308. [CrossRef]
5. Soutis, C. Fibre reinforced composites in aircraft construction. *Prog. Aerosp. Sci.* **2005**, *41*, 143–151. [CrossRef]
6. Giasin, K.; Ayvar-Soberanis, S. An investigation of burrs, chip formation, hole size, circularity and delamination during drilling operation of GLARE using ANOVA. *Compos. Struct.* **2017**, *159*, 745–760. [CrossRef]
7. Teti, R. Machining of composite materials. *CIRP Ann.-Manuf. Technol.* **2002**, *51*, 611–634. [CrossRef]
8. Che, D.; Saxena, I.; Han, P.; Guo, P.; Ehmann, K. Machining of carbon fiber reinforced plastics/polymers: A literature review. *J. Manuf. Sci. Eng.-Trans. ASME* **2014**, *136*, 034001. [CrossRef]

9. Xu, J.; Li, C.; Mi, S.; An, Q.; Chen, M. Study of drilling-induced defects for CFRP composites using new criteria. *Compos. Struct* **2018**, *201*, 1076–1087. [CrossRef]
10. Xu, J.; Li, C.; Chen, M.; El Mansori, M.; Ren, F. An investigation of drilling high-strength CFRP composites using specialized drills. *Int. J. Adv. Manuf. Technol.* **2019**, *103*, 3425–3442. [CrossRef]
11. Wang, D.H.; Ramulu, M.; Arola, D. Orthogonal cutting mechanisms of graphite/epoxy composite. Part II: Multi-directional laminate. *Int. J. Mach. Tools Manuf.* **1995**, *35*, 1639–1648. [CrossRef]
12. Wang, D.H.; Ramulu, M.; Arola, D. Orthogonal cutting mechanisms of graphite/epoxy composite. Part I: Unidirectional laminate *Int. J. Mach. Tools Manuf.* **1995**, *35*, 1623–1638. [CrossRef]
13. Iliescu, D.; Gehin, D.; Iordanoff, I.; Girotabc, F.; Gutiérrez, M.E. A discrete element method for the simulation of CFRP cutting *Compos. Sci. Technol.* **2010**, *70*, 73–80. [CrossRef]
14. Yan, X.; Reiner, J.; Bacca, M.; Altintas, Y.; Vaziri, R. A study of energy dissipating mechanisms in orthogonal cutting of UD-CFRP composites. *Compos. Struct.* **2019**, *220*, 460–472. [CrossRef]
15. Wang, C.; Chen, Y.; An, Q.; Cai, X.; Ming, W.; Chen, M. Drilling temperature and hole quality in drilling of CFRP/aluminum stacks using diamond coated drill. *Int. J. Precis. Eng. Manuf.* **2015**, *16*, 1689–1697. [CrossRef]
16. Wang, H.X.; Zhang, X.H.; Duan, Y.G. Effects of drilling area temperature on drilling of carbon fiber reinforced polymer composites due to temperature-dependent properties. *Int. J. Adv. Manuf. Technol.* **2018**, *96*, 2943–2951. [CrossRef]
17. Merino-Pérez, J.L.; Royer, R.; Ayvar-Soberanis, S.; Merson, E.; Hodzic, A. On the temperatures developed in CFRP drilling using uncoated WC-Co tools Part I: Workpiece constituents, cutting speed and heat dissipation. *Compos. Struct.* **2015**, *123*, 161–168 [CrossRef]
18. Merino-Pérez, J.L.; Hodzic, A.; Merson, E.; Merson, E.; Hodzic, A. On the temperatures developed in CFRP drilling using uncoated WC-Co tools Part II: Nanomechanical study of thermally aged CFRP composites. *Compos. Struct.* **2015**, *123*, 30–34 [CrossRef]
19. Rubio, J.C.C.; Panzera, T.H.; Scarpa, F. Machining behaviour of three high-performance engineering plastics. *Proc. Inst. Mech Eng. Pt. B-J. Eng. Manuf.* **2015**, *229*, 28–37. [CrossRef]
20. Wang, B.; Gao, H.; Cao, B.; Yuan, Z.; Zhe, Z. Mechanism of damage generation during drilling of carbon/epoxy composites and titanium alloy stacks. *Proc. Inst. Mech. Eng. Pt. B-J. Eng. Manuf.* **2014**, *228*, 698–706. [CrossRef]
21. Davim, J.P.; Reis, P. Drilling carbon fiber reinforced plastics manufactured by autoclave-experimental and statistical study. *Mater Des.* **2003**, *24*, 315–324. [CrossRef]
22. Davim, J.P.; Reis, P. Study of delamination in drilling carbon fiber reinforced plastics (CFRP) using design experiments. *Compos Struct.* **2003**, *59*, 481–487. [CrossRef]
23. Xu, J.; An, Q.; Cai, X.; Chen, M. Drilling machinability evaluation on new developed high-strength T800S/250F CFRP laminates *Int. J. Precis. Eng. Manuf.* **2013**, *14*, 1687–1696. [CrossRef]
24. Bonnet, C.; Poulachon, G.; Rech, J.; Girard, Y.; Costes, J.P. CFRP drilling: Fundamental study of local feed force and consequences on hole exit damage. *Int. J. Mach. Tools Manuf.* **2015**, *94*, 57–64. [CrossRef]
25. Babu, J.; Sunny, T.; Paul, N.A.; Mohan, K.P.; Philip, J.; Davim, J.P. Assessment of delamination in composite materials: A review *Proc. Inst. Mech. Eng. Pt. B-J. Eng. Manuf.* **2016**, *230*, 1990–2003. [CrossRef]
26. Xu, J.; An, Q.; Chen, M. An experimental investigation on cutting-induced damage when drilling high-strength T800S/250F carbon fiber-reinforced polymer. *Proc. Inst. Mech. Eng. Pt. B-J. Eng. Manuf.* **2017**, *231*, 1931–1940. [CrossRef]
27. Xu, J.; Huang, X.; Chen, M.; Davim, J.P. Drilling characteristics of carbon/epoxy and carbon/polyimide composites. *Mater. Manuf. Processes* **2020**, *35*, 1732–1740. [CrossRef]
28. Su, F.; Wang, Z.; Yuan, J.; Cheng, Y. Study of thrust forces and delamination in drilling carbon-reinforced plastics (CFRPs) using a tapered drill-reamer. *Int. J. Adv. Manuf. Technol.* **2015**, *80*, 1457–1469. [CrossRef]
29. Ameur, M.F.; Habak, M.; Kenane, M.; Aouici, H.; Cheikh, H. Machinability analysis of dry drilling of carbon/epoxy composites, cases of exit delamination and cylindricity error. *Int. J. Adv. Manuf. Technol.* **2017**, *88*, 2557–2571. [CrossRef]
30. Kuo, C.; Wang, C.; Ko, S. Wear behaviour of CVD diamond-coated tools in the drilling of woven CFRP composites. *Wear* **2018**, *398–399*, 1–12. [CrossRef]
31. Rawat, S.; Attia, H. Wear mechanisms and tool life management of WC-Co drills during dry high speed drilling of woven carbon fibre composites. *Wear* **2009**, *267*, 1022–1030. [CrossRef]
32. Faraz, A.; Biermann, D.; Weinert, K. Cutting edge rounding, an innovative tool wear criterion in drilling CFRP composite laminates. *Int. J. Mach. Tools Manuf.* **2009**, *49*, 1185–1196. [CrossRef]
33. Wang, X.; Kwon, P.Y.; Sturtevant, C.; Kim, D.; Lantripc, J. Tool wear of coated drills in drilling CFRP. *J. Manuf. Process.* **2013**, *15*, 127–135. [CrossRef]
34. Ismail, S.; Dhakal, H.; Dimla, E.; Popov, I. Recent advances in twist drill design for composite machining, A critical review. *Proc. Inst. Mech. Eng. Pt. B-J. Eng. Manuf.* **2017**, *231*, 2527–2542. [CrossRef]
35. Fu, R.; Jia, Z.Y.; Wang, F.J.; Jin, Y.; Sun, D.; Yang, L.J.; Cheng, D. Drill-exit temperature characteristics in drilling of UD and MD CFRP composites based on infrared thermography. *Int. J. Mach. Tools Manuf.* **2018**, *135*, 24–37. [CrossRef]

6. Kubher, S.; Gururaja, S.; Zitoune, R. In-situ cutting temperature and machining force measurements during conventional drilling of carbon fiber polymer composite laminates. *J. Compos. Mater.* **2021**, *55*, 2807–2822. [CrossRef]
7. Xu, J.; Li, C.; Chen, M.; El Mansori, M.; Davim, J.P. On the analysis of temperatures, surface morphologies and tool wear in drilling CFRP/Ti6Al4V stacks under different cutting sequence strategies. *Compos. Struct.* **2020**, *234*, 111708. [CrossRef]

Journal of
Composites Science

MDPI

Article

Experimental Investigation on Machine-Induced Damages during the Milling Test of Graphene/Carbon Incorporated Thermoset Polymer Nanocomposites

Jogendra Kumar [1], Kumar Abhishek [2], Jinyang Xu [3] and Rajesh Kumar Verma [1,*]

[1] Materials and Morphology Laboratory, Department of Mechanical Engineering, Madan Mohan Malaviya University Technology, Gorakhpur 273010, Uttar Pradesh, India; jkme@mmmut.ac.in
[2] Department of Mechanical and Aero-Space Engineering, Institute of Infrastructure, Technology, Research and Management (IITRAM), Ahmedabad 380026, Gujarat, India; krabhishek1987@gmail.com
[3] State Key Laboratory of Mechanical System and Vibration, School of Mechanical Engineering, Shanghai Jiao Tong University, Shanghai 200240, China; xujinyang@sjtu.edu.cn
* Correspondence: rkvme@mmmut.ac.in; Tel.: +91-8400444068

Abstract: The fiber laminate composites are extensively used in aerospace, aircraft, automotive components due to their high stiffness, corrosion, moisture resistance, low weight, and durability features. These fiber composites are modified with nanomaterials to acquire the desired manufacturing properties. The complex structure and anisotropic features differ from metals and their alloys Additionally, the machining principles of fiber laminates significantly differ from conventional engineering materials. The present work investigates the machining behavior and permeates the damage generated while milling of graphene-modified carbon-fiber reinforced polymer nanocomposites (G/C@FRNC). The surface damages and defects caused in the milling samples have been examined through the high-resolution spectroscopy test. The influence of machining constraints such as cutting speed (N), feed rate (F), depth of cut (D), and graphene oxide weight % (GO) has been investigated to achieve the desired milling performances viz. material removal rate (MRR), cutting force (Fc), surface roughness (Ra), and delamination factor (F_d). The outcomes indicated that the cutting parameters and graphene nanomaterial prominently affects the milling responses. The addition of graphene improves the machinability of proposed nanocomposites with lesser defects generated. However, its higher addition can lead to the phenomenon of agglomeration that can reduce the machining efficiency. The damages and delamination generated in the machined sample are low at a higher cutting speed. This work suggests a new system to control the damage and defects to enhance the laminate samples' quality and productivity.

Keywords: graphene oxide; polymer; composite; milling; carbon fiber

Citation: Kumar, J.; Abhishek, K.; Xu, J.; Verma, R.K. Experimental Investigation on Machine-Induced Damages during the Milling Test of Graphene/Carbon Incorporated Thermoset Polymer Nanocomposites. *J. Compos. Sci.* **2022**, *6*, 77. https://doi.org/10.3390/jcs6030077

Academic Editor: Francesco Tornabene

Received: 27 January 2022
Accepted: 28 February 2022
Published: 2 March 2022

1. Introduction

For the last two decades, fiber laminate composites have been highly utilized in the components of aerospace, naval, space, and automotive industries with better mechanical properties and modified fatigue life [1]. It becomes an adequate substitute for traditional engineering materials such as metallic alloys and non-metallic materials. The selected manufacturing materials must have unique physical and structural properties with a combination of low specific weight and high resistance to degradation. Under these multifunctional conditions, it can ensure economic performance and safety factors. In addition to thermoset polymers, cross-linked connection contributes to higher stiffness, thermal and mechanical properties. The cross-linked nature, anisotropic, and non-homogeneity of fiber composite creates challenging machining issues [2,3]. In the epoxy resin, nanofiller materials reinforcing were regarded as an effective technique to boost the strength and stiffness under different loading conditions. In addition to the feasible nanomaterial in the epoxy matrix could significantly improve the flexural and tensile behavior [4,5]. Various carbon nanomaterials (CNMs), such as fullerenes, nanotubes (NT), nanorods (NR), graphene derivatives, etc., have been used to enrich polymer composite mechanical behavior [6]. The mechanical properties of fullerene (C60) epoxy nanocomposites were characterized by Rafie et al. [7] at the different aspect ratio of fullerene nanoparticles in the polymer matrix. At relatively low nanofiller loading fractions, the epoxy (resin) mechanical characteristics improved substantially through the fullerene nanoparticle (0.1 to 1% of the resin). The effect of multi-walled carbon nanotubes (MWCNTs) on polymer nanocomposites' impact and bending properties at room temperature was examined by Liang et al. [8]. The study disclosed the nanomaterial weight fraction could increase it. The same findings (flexural strength) were explored by Gantayat et al. [9] to analyze the reinforced hybrid composite. Watson et al. [10] researched a modified dispersion method of graphene oxide using a sonic process. The vacuum resin infusion for nanocomposites reinforced by carbon fiber is performed for the fabrication phase. The research explored the effects of the developed carbon nanocomposites' tensile strength and flexural strength. It was noted that the tensile strength of the neat epoxy fiber composite was lower than the GO-modified composites. A similar trend was observed by Cho et al. [11] The positive impacts of graphene nanoparticles (GO) on epoxy resin-based composites were analyzed by Abdullah et al. [12] The GO/epoxy composite was cast and prepared at room temperature by the casting process. The finding decided that nanomaterial performs a vital position in enhancing mechanical properties such as tensile strength, impact strength, and hardenability. The morphological surface analysis confirmed that cracks in the composite were prevented from propagating. Additionally, the results claimed that a variety of industrial applications might use GO/epoxy composites. To produce highly conductive textiles with mechanically tunable, hydrophobic, and TASER protective properties, Ghosh et al. [13] demonstrated an easy-to-scale method of dipping and drying materials. Protective EM-treated organic wool composite smart textile with sensitive touch switches for household and wireless connectivity. With the help of the rGO nanosheets, a robust electrical and EMI shielding network was constructed in electrical and electromagnetic interference protecting application areas. Additionally, Ghosh et al. [14] developed silver nanoparticle-decorated graphene sheets (rGO/Ag) using a two-stage wet mixing technique. Using non-ionic polymer adhesive to create conductive coatings protects against radiation pollution from electronic technologies and equipment.

Milling is the primary machining process in industries to create the slots and channels for the assembly of products. For several years, the milling of isotropic metallic materials has been studied in depth, but these results can sometimes not apply to the machining of polymer laminate composites. According to the work of Davim et al. [15], delamination and fiber damages are generated in CFRP composites during machining in a significant way. It affects the machining efficacy and quality of the samples. It occurs in laminate polymer due to the anisotropic and non-homogeneous nature and similar remarks were observed in the finding of eminent scholars [16–18]. The macro-reinforced (fiber) composites' existing properties are prominently improved by supplementing nanomaterials. Doping theory is sometimes used in fiber composites to investigate polymers' physical and mechanical aspects. However, several valuable studies examined the cutting tool geometry constraints and tool materials for orthogonal cutting [19,20]. The fiber orientation, fiber size, and weave design also affect the machining performances [18,21,22].

The pioneer scholars limitedly attempt the studies on the use of graphene oxide in carbon/epoxy composites. It requires more attention to efficiently utilize polymer nanocomposites for multifunctional products. Studies show that graphene oxide has exceptional characteristics due to better mechanical properties, aspect ratio, and dispersion [5,23,24]. However, machining efficiency computations and control of damages are highly required in the polymer manufacturing sector to improve product quality and productivity.

From an exhaustive state of the art, it has been remarked that the machinability evaluation of laminate composites modified by carbon nanomaterials is passing through the preliminary phase. Researchers' attention, practicing engineers, and academic interest are needed to overcome these modified polymers' machining drawbacks and limitations. It can become an emerging area of research for the stakeholder to create cost-effective and stable products. Because of the emerging mechanical properties, the present work investigates the damages and defects generated during the milling of graphene-modified carbon-fiber reinforced polymer nanocomposites (G/C@FRNC). The control of process parameters, such as the cutting speed (N), feed rate (F), depth of cut (D) and graphene oxide weight % (GO%), is proposed to overcome the surface defects and damages that ensue during the milling procedure. These defects hamper the efficiency of product development by decreasing structural rigidity. The machined surface is examined with microstructural analysis, and the impact of varying constraints has been evaluated. The milling performance is estimated for the desired value of MRR, F_c, R_a and F_d through the control of process constraints.

The present work shows the defect's spectroscopy investigation on the slot wall of the unidirectional laminates composite. The issue of deterioration directly contributes to the breakage of fibers and occurs through the bending and shearing of carbon fibers. The outcomes of the literature work have demonstrated an improvement in the fundamental and functional properties of hybrid nanocomposites. This article proposed a pioneering method to suppress the delamination and damages versus machining parameters.

2. Experimental Work

2.1. Development of Graphene-Modified Carbon-Fiber Reinforced Polymer Nanocomposites

A matrix (epoxy 520) and straight woven 400 GSM carbon fibers were used for the nanocomposite fabrication. The graphene oxide with 255 m^2/g surface area > 99% carbon impurity is used. Due to epoxy's high viscosity, it is challenging to make a homogeneous mixture of graphene nanoparticles and epoxy. Epoxy (thermoset resin at higher viscosity value 1.162 gm/cm^3) was stirred with different amounts of dispersed graphene nanoparticle (1, 2, and 3%) up to 30 min at 60 °C. Finally, a binding agent (Hardener-D) was introduced with the ratio of 10:1 and stirred again with a temperature of 27 °C. The fabrication procedure of nanocomposite samples is described in the schematic diagram (Figure 1). For control of agglomeration level in the mixture, a balance addition of graphene nanomaterial (1, 2, and 3% weight ratio) is employed with reference to previous experimental results and theoretical study [25–27]. Three samples with a selected set of milling parameters were assessed to enhance the reliability of the experimental results. The Hand lay-up method prepares the laminate composite of 18 layers of woven fiber. For better cutting force analysis (in depth), dimensions of 100 mm in length, 100 mm in width, and 10 mm in thickness have been selected for the fabricated sample.

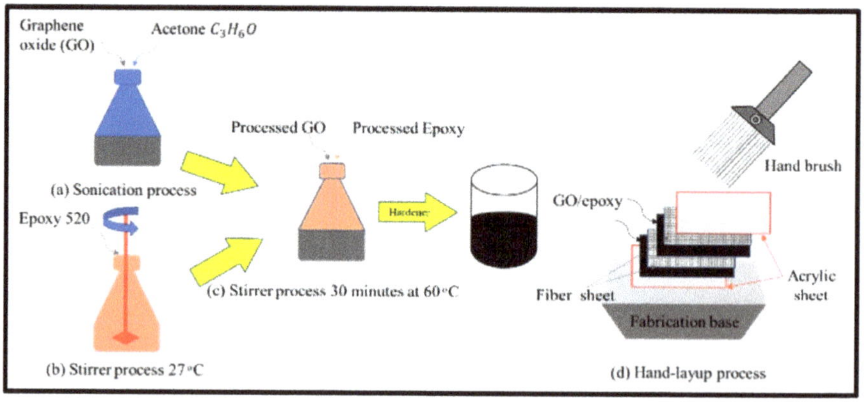

Figure 1. Nanocomposite fabrication procedure (**a**) Sonication process, (**b**) Stirrer process 27 °C, (**c**) Stirrer process 60 °C, (**d**) Hand-layup process.

Fiber thickness 400 GSM = 0.45 mm
Number of layers used in the development composite = 18 layers
One-layer area = 250×250 mm^2
One-layer weight = $\frac{250 \times 250}{400}$ = 25 g
Eighteen-layer weight = $25 \times 18 = 450$ g
Fiber: resin = 70:30
$\frac{450}{R} = \frac{70}{30}$
R = 192.85 g
1 wt.% of GO = 1.9285 g
2 wt.% of GO = 3.875 g
3 wt.% of GO = 5.7855 g
Epoxy: Hardener = 10:1
$\frac{192.85}{H} = \frac{10}{1}$
H = 19.285 g

GO/C@FRNC laminates nanocomposite with 10 mm thickness (18 plies) were examined in this study. The detailed mechanical properties of GO/C@FRNC are described in Table 1.

Table 1. Mechanical properties of G/C@FRNC laminates.

Sample	Tensile Strength (MPa)	Flexural Strength (MPa)
1	934.35	442.55
2	875.04	363.70
3	774.01	289.95

2.2. Experimental Setup and Response Measurement during Milling of Modified Composites

The fiber polymer nanocomposite plate was machined (milling) under dry environments to investigate the cutting process. A vertical milling machine computerized controlled setup (Model No. MBV 35 TC20) was employed for the milling test of composite samples. The milling cutter tool is made of SiC coated TiAlN, with a 5 mm diameter is selected for the experiment. A dynamometer (Model No. MLB-PML-300) for cutting force and precision balance meter weight was used for MRR calculation during experimentation. The surface roughness tester (Mitutoyo Model No. SJ 210) was measured using mathematical average at three different slot locations, as shown in Figure 2.

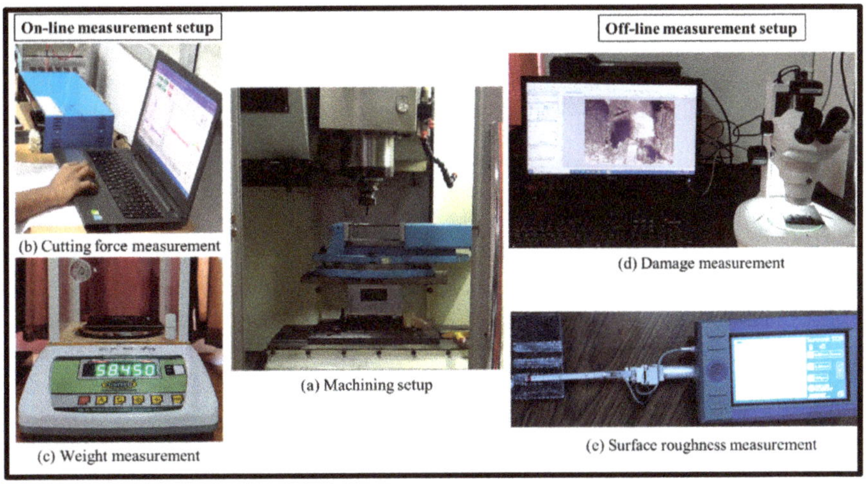

Figure 2. Experimentation setup. (**a**) Machining setup, (**b**) cutting force measurement, (**c**) weight measurement, (**d**) damage measurement and (**e**) surface roughness measurement.

The tomography microscope (Model No. ZSM3780 T2) has software ULTRA-CMOS510 to estimate delamination value. The milling parameters are described in Table 2. The milling delamination factor (F_d) characterizes the machined slot delamination region (milling slot) F_d is expressed as the ratio of the maximum damage slot length of the machined zone to the nominal slot diameter. The F_d was computed by using the below formula:

$$Delamination\ factor\ (F_d) = \frac{Maximum\ damage\ slot\ width\ (W_{max})}{Nominal\ damage\ slot\ width\ (W)} \quad (1)$$

Table 2. Process parameters and their working range.

Sr. No.	Factors	Nomenclature	Range/Unit
1	Cutting speed	N	12.56–37.68 m/min
2	Feed rate	F	80–240 mm/min
3	Depth of cut	D	0.5–1.5 mm
4	Graphene nanomaterial	G	1–3 wt.%

3. Result and Discussion

3.1. Influence of Process Parameters on MRR

The cutting speed increased noticeably at a control material removal rate, which is contradictorily associated with traditional machining. The material removal rate (MRR) showed a higher value at medium cutting speed (25.12 m/min). The higher feed value is preferred for MRR; however, its higher level would be more robust. The predominant MRR decreased at a lower cutting depth compared with the conventional materials. In contrast, the cutting depth changed at a higher value, increasing the rate of material removal similar to that of the feed. If the cutting is more thorough, the workpiece's fibers have been cut entirely, and less impact is put on the tool, and successful material removal could be achieved [28]. In turn, results in lower specific cutting pressure, and therefore, high cutting depth to achieve the best performance characteristics is preferred [28]. These may indicate alterations due to irregularities in cutting speed, depending on the cutting force's parameters. In representing the cutting force properties, epoxy/CFRP modifiers are likely to be more ductile because of increased temperature and greater molecular mobility than traditional materials at higher feed speeds. While the lack of effective heat loss from the machined surface and the composite chip accelerates tool wear. At this time, the material's low thermal conductivity contributes to the difficulties of the milling operation. Figure 3 demonstrates that with higher cutting depth (1.5 mm), feed rate (240 mm/min), medium cutting speed (25.12 m/min), and wt.% of graphene nano content (2%), the MRR substantially improved the machining process.

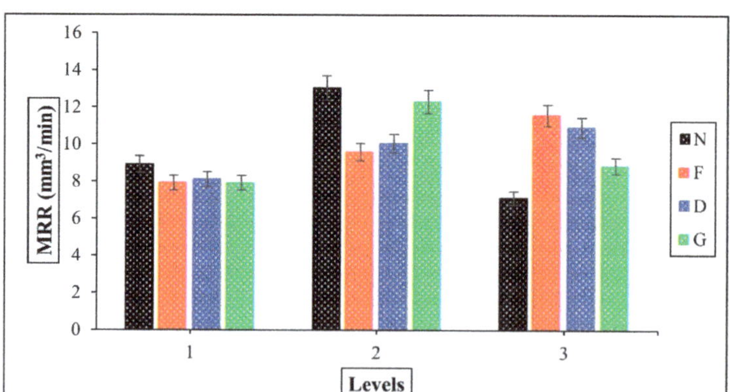

Figure 3. MRR vs. milling factors analysis.

3.2. Influence of Process Parameters on Cutting Force

Cutting forces are the primary factor affecting the quality of the machined surface. It greatly reduces generically with rising cutting speeds (N) and a control combination of the cutting depth and feed rate (F). There is variation in CFRP's cutting force conditions due to the fiber/polymeric material's anisotropy nature at different feed rates. At various cutting speeds (low, medium, and higher), the modified epoxy/CFRP composite material displayed contrasting cutting force features. The hybrid materials exhibit brittle fracture behavior at low cutting speeds that affects their machinability, but medium cutting speed influences the process. The fiber experiences a high strain rate under a high cutting speed, causing it to fail at a lower strain or in a brittle form [29]. It raises the strain rate leading to low-stress fractures and material degradation, thereby increasing the cutting region's temperature and improving the composite's molecular ductility. During machining, a large material flow was observed with the cut fibers compared to higher cutting speed to medium cutting speed, so moderate speed is recommended for lower cutting force [28]. Figure 4 displays the graphical representation of cutting forces versus process parameters at various cutting speeds, feed rates, cutting depths, and weight percent of the graphene filler, respectively. The cutting force is found to a minimum with the cutting speed up to 25.12 m/min and 160 mm/min feed. The reduction force is the minimum due to the ploughing impact at a high feed rate [28]. The cutting force is substantially lowest at a lower cut depth and weight % of graphene nanomaterial. Another remarkable aspect seems to be that the cutting force's intensity increased dramatically for higher cutting depth. It is possible due to the rate at which the cutting tool penetrates more with increased cutting depth and extracts more material, resulting in increased cutting force [30].

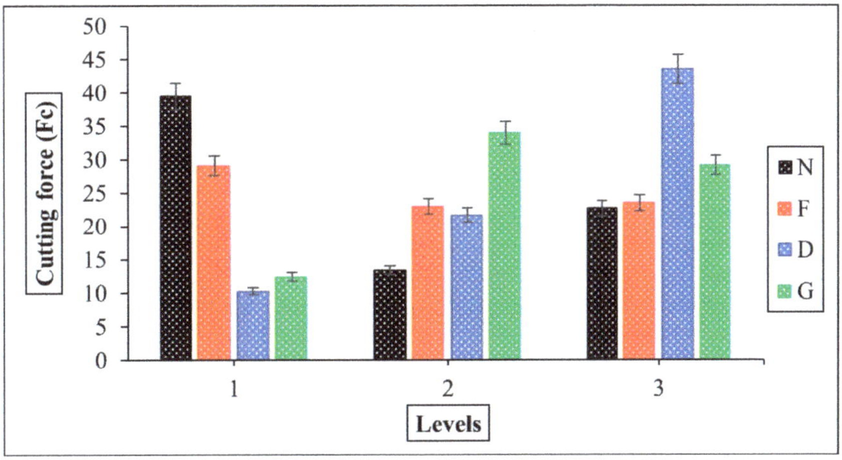

Figure 4. Cutting force vs. milling factors analysis here.

3.3. Influence of Process Parameters on Surface Roughness

Unlike in the traditional milling process where cutting is speedier, feed are more significant, machining surfaces of modified epoxy/CFRP. The explanation for the simultaneous presence of fibre and resin interfacial debonding and the weak resin solidification at high temperatures. Several investigations demonstrated that the polymer is softened by increasing the cutting temperature during the milling process. The degradation of temperatures above the matrix resin can occur, contributing to negative processing quality. The effect of this phenomenon is a poor resin support function for carbon fibers [31]. Alongside that, slightly higher surface roughness can be observed at a high feed level. In the machine tool, the feed rates and cut depth cause vibrations that influence the performance and surface quality of the milling. The reason might be that the rise in cutting force and higher tool wear also occurred with increased depth of cut and defected surface quality [30]. In polymeric materials, lower feed rates related to the material and composites' inhomogeneity create more vibrations during the cutting process. Figure 5 indicates the plots of surface roughness over different levels of process parameters.

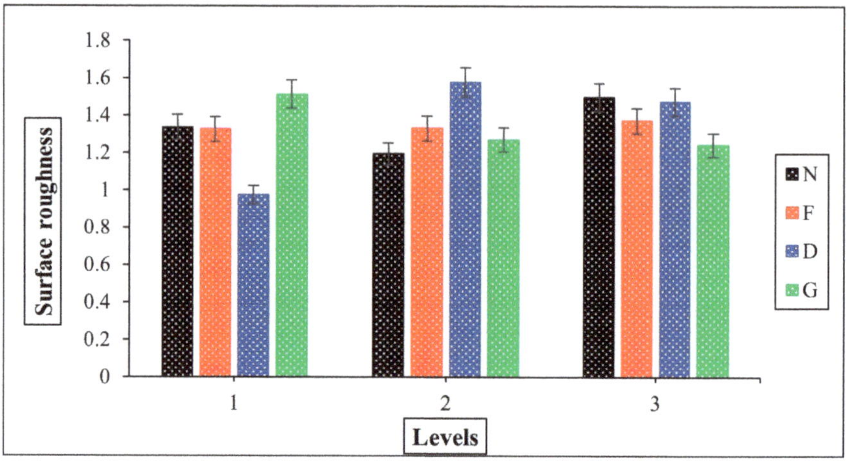

Figure 5. Surface roughness vs. milling factors analysis.

A smooth surface finish was remarked at 25.12 m/min cutting speed with a lower depth of cut. For minimal surface roughness in FRP composites' processing, medium cutting and lower feed rates are recommended [32]. The observation shows that all three weight % of graphene nanomaterials are impacted with a high aspect ratio. The tool-workpiece interface tends to communicate with lubricants due to graphene nanomaterial presence. It indicates the prevalence of ductile transformation in the polymer increases for improved machinability. However, cutting speed is consistent with graphene nanomaterial and is dominated by modified epoxy/CFRP soft machining. With increased nano content, the decrease in surface roughness through 25.12 m/min is indicative of the need to mill these materials at an improved surface finish. Figure 5 shows that with a cutting speed of 25.12 m/min, the surface roughness decreases significantly at lower cutting depth, feed rate, and higher graphene content of wt.%.

The machined sample was observed with a microscope image at different cutting conditions to demonstrate the fiber or composite interfaces' micro deterioration at different milling operation levels. Figures 6–8 show the typical damages incurred while conducting the machining trials. The microscope image of the fiber pullout, uncut fiber, fiber fraying, feed mark, cavity hole, fiber fracture, and crack regarding the different cutting speed conditions indicates higher surface roughness during the milling experiments. The appearance of surface damages may be because of high cutting speed and high feed level. Additionally, it leads to material defect encountered during the composite material process. The microscope images of the slot surfaces are shown in Figure 6a–c. It is observed that surface cavity, feed marks and fiber pullout damage have occurred at the slot region. This is due to interfacial debonding of fibers, higher feed, and the poor support of fiber. Medium cutting speed is recommended for minimal surface roughness in the machining of FRP composites [32]. From Figure 7a–c shows the very few cavity forms, fiber breakage, and pullout. Figure 8a–c shows that more surface damages occurred, propagating the cracks in the laminates. However, high cutting speed leads to higher breakage, pullout, serious cavity formation, and brittle fracture resulting in poor surface finish. This is because the resin is softened with increased cutting temperature, higher feed, and nanomaterials agglomeration.

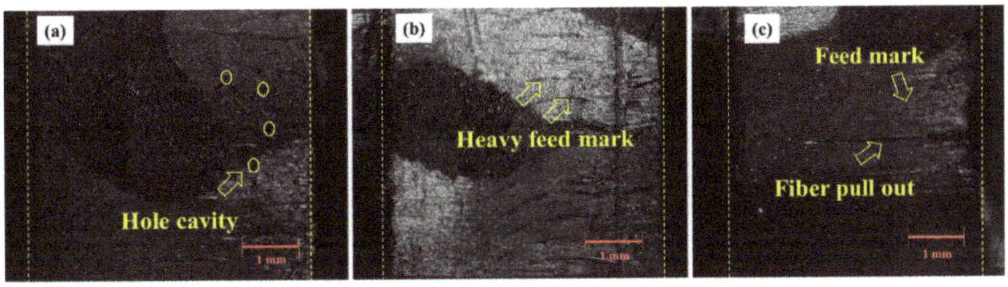

Figure 6. Microscope image of the machined slot at lower cutting speed for (**a**) 1 wt.% GO, (**b**) 2 wt.% GO, and (**c**) 3 wt.% GO.

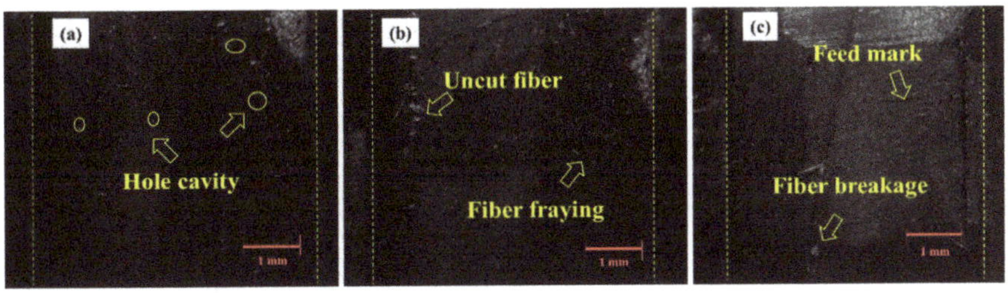

Figure 7. Microscope image of the machined slot at medium cutting speed for (**a**) 1 wt.% GO, (**b**) 2 wt.% GO, and (**c**) 3 wt.% GO.

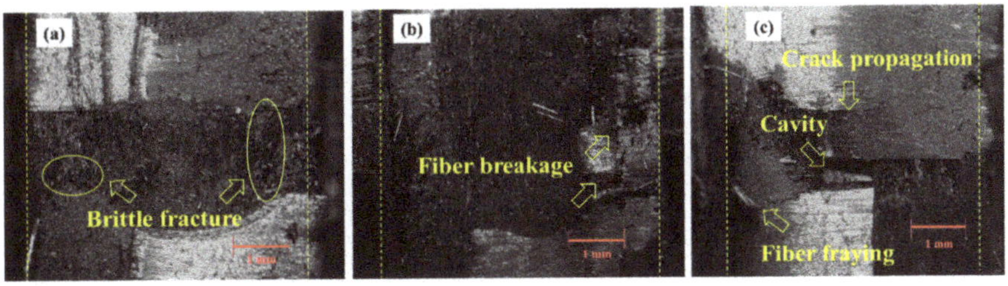

Figure 8. Microscope image of the machined slot at higher cutting speed for (**a**) 1 wt.% GO, (**b**) 2 wt.% GO, and (**c**) 3 wt.% GO.

3.4. Influence of Process Parameters on Milling-Induced Delamination

During the milling of fiber laminates, delamination is a severe and challenging machining issue in polymer composites. It affects product assembly, productivity, and the overall efficiency of the manufacturing process. The control of cutting conditions can overcome this machining-induced damage. The cutting force estimated through the computerized toolbox dynamometer is peaked when the maximum force occurred during the machining process. In ceramics, metallic chip formation is a vital source of information on the milling process's deformation characteristics, besides composites where fiber and resin have unique physical properties and perforation and damage analysis parameters are different. In comparison, the critical cutting force of the synergistic reinforced composite of graphene oxide was greater. The delamination region in the milling test of the proposed nanocomposite becomes more complex due to the plain-woven fiber properties and weaves. These properties cause variation in the cutting force. The high strain rates lead to chip fragmentation, and feed rates indicate a resin fracture early due to the modified epoxy/CFRP milling process combined with high cutting speed. However, the heat produced can increase the cutting zone's temperature with a higher combination of feed and cutting speed, increasing the material's molecular chains' long-range mobility, thus increasing its ductility. In this way, the fracture of fibers tends to increase, causing more severe damage [29]. At the higher cutting speed and feed rate, the chip microstructure is prominent in the long continuous carbon fiber strands' instantaneous fracturing, demonstrating substantial fiber crack propagation on the chip at intermittent lengths. It was found (Figure 9) that due to high rubbing at the work–tool interface at high cutting speed and feed, the bond strength between fiber and matrix decreased so that fibers are either peeling (delamination) or removed from the matrix.

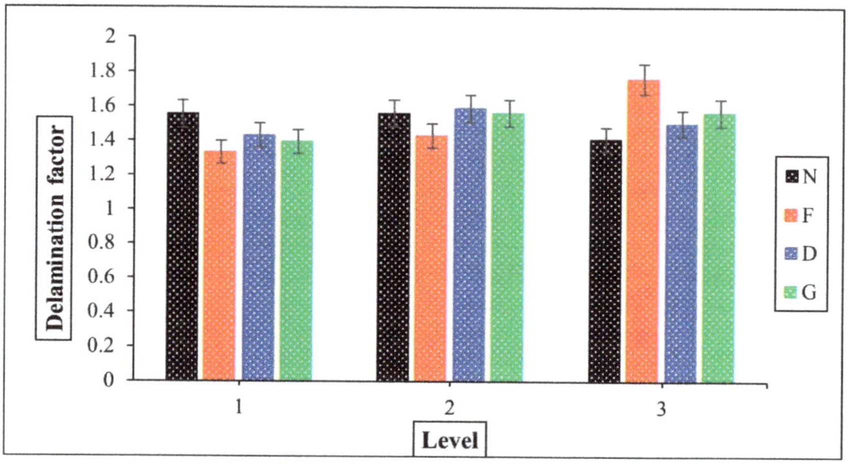

Figure 9. Delamination factor vs. milling factors analysis.

The machined sample microstructure images reveal serrations and segmentation with higher surface damage at the highest feed rate (240 mm/min). A high feed rate leads to more significant vibration during the cutting process, contributing to fracture failure. The higher feed rates resulted in damage due to lower interfacial strength caused by the breakage of the fiber/matrix [15]. The segmentation pattern is sufficient at the lower feed and higher cutting speed, indicating less failure, better machinability, and smooth finishing. At a higher cutting speed, a few polymeric resin crack propagations with fiber pullout is observed at a slightly lower feed rate. This would be due to the lack of bending of the cutting tool, which is accomplished by decreasing the cutting force, smooth discharge of the chip when the cutting speed of the cutting chips exceeds the cutting speed; and a decrease in uncut fiber yarn burrs [19,33]. However, with increased depth of cut, extensive machining surface damage, higher fiber pull can be confirmed of epoxy/CFRP composite [29]. In addition, the damage caused by improved mechanical properties and microstructure is diminished by loaded nanofiller samples [34]. Therefore, graphene nanomaterial affected the cutting force, contributing to more uniform surface topography in doped CFRPs without substantial fibers' breakage [35]. Carbon nanomaterials improve their high surface-to-volume ratio to the contact surface between the reinforcement and the matrix, which results in greater strain distribution and damage resistance [36]. As previously demonstrated by similar research, the G/carbon-fiber-reinforced composites can achieve less delamination damage with a lower feed rate [36,37]. Figure 10a–f show the machining effect on modified epoxy of carbon nanocomposite with 1–3 wt.% of graphene oxide. The aspect of nanomaterials the influence of machine composite is revealed with three different samples. Figure 10a sample containing 1 wt.% of nanomaterial with a 20 µm scale with minimal fiber debonding. Figure 10b sample containing 2 wt.% of nanomaterial with a 20 µm scale with micro-crack, crack propagation, and machined marks. Figure 10c sample containing 3 wt.% of nanomaterial with a 20 µm scale with severe breakage of the machined sample fiber. The current study results were compared to previous ones, and it was determined that the method was feasible in machining while responses were taken individually (Table 3).

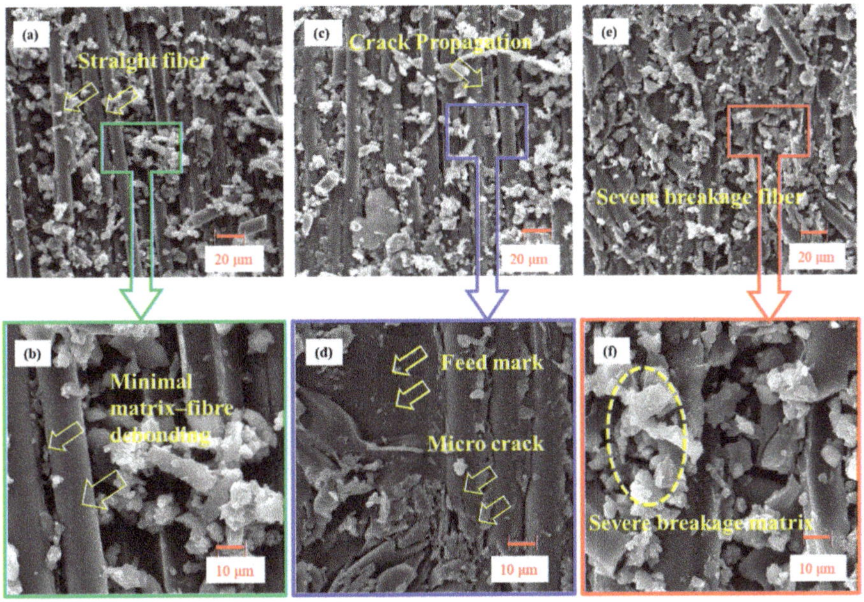

Figure 10. Spectroscopy analysis (SEM) of milling workpiece for (**a**) 1 wt.% GO at 20 μm, (**b**) 1 wt.% GO at 10 μm, (**c**) 2 wt.% GO at 20 μm, (**d**) 2 wt.% GO at 10 μm, (**e**) 3 wt.% GO at 20 μm, and (**f**) 3 wt.% GO at 10 μm .

Table 3. Comparative analysis with the previous result.

Sr. No.	Condition	Response	Present Study	Previous Study	Error%	Ref.
1	N2F3D3G2	MRR	18.768	17.0484	9.16%	
2	N2F2D1G1	Fc	4.706	4.706	-	[38]
3	N2F2D1G3	Ra	0.716	0.730	1.95%	
4	N3F1D1G1	Fd	1.043	1.043	-	[39]

4. Conclusions

The works focus on improving the quality and productivity concerns during the milling of graphene-modified carbon-fiber reinforced polymer nanocomposites (G/C@FRNC). The microstructural investigation through high-resolution results demonstrates the milling efficiency of the fabricated nanocomposites. The effect of cutting parameters on the MRR, Fc, Ra, and Fd has been explored to estimate the polymers' machinability features. The findings of the present article are as follow:

- The cutting force and MRR trend observed in the proposed laminate nanocomposite are similar to traditional materials. The value of MRR and Fc is higher at the combined effect of the higher feed and medium cutting speed;
- The quality of the machined surface can be controlled at a medium cutting speed, lower feed, lower depth of cut and a higher addition of nanomaterial;
- A significant amount of matrix (epoxy) damage is observed at lower feed rates and higher cutting speeds. Furthermore, chip separation around the edge of the machined sample (milling slot) was observed at lower cutting speed and higher value feed;
- The lower value of cutting speed and higher feed value was observed for fiber fraying, fiber pullout, and matrix smearing, with fractured fibers (damage) firmly embedded

in the matrix. However, medium cutting speed machining can be used to effectively machine modified epoxy/CFRP laminates at optimized lower feed rates, the depth of cut depth, and weight% of graphene nanoparticles;
- The addition of graphene oxide contributes significantly to the reduction of milling induce damages. A little addition of graphene improves the machining efficiency of the proposed nanocomposite.

The outcomes of the works show that the proposed nanocomposites can withstand structural component needs. Additionally, the machined surface quality at a medium cutting speed shows the higher application potential for an effectual milling test. The inclusion of other machining operations, such as drilling, turning, etc., can be used as the scope of future work.

Author Contributions: J.K.: Writing—original draft, Methodology, Conceptualization. K.A.: Writing–review & editing, formal analysis. J.X.: Writing—review & editing, formal analysis. R.K.V.: Writing—review & editing, Software, Supervision, Validation, Conceptualization. All authors have read and agreed to the published version of the manuscript.

Funding: This research did not receive any specific grant from funding agencies in the public, commercial, or not-for-profit sectors.

Acknowledgments: The authors are very grateful to the M.M.M University of Technology, India for continuous support during this work.

Conflicts of Interest: The authors declared no potential conflict of interest.

References

1. Wiegmann, E.; Helmers, L.; Michalowski, P.; Kwade, A. Highly scalable and solvent-free fabrication of a solid polymer electrolyte separator via film casting technology. *Adv. Ind. Manuf. Eng.* **2022**, *3*, 100065. [CrossRef]
2. Altin Karataş, M.; Gökkaya, H. A review on machinability of carbon fiber reinforced polymer (CFRP) and glass fiber reinforced polymer (GFRP) composite materials. *Def. Technol.* **2018**, *14*, 318–326. [CrossRef]
3. Geng, D.; Liu, Y.; Shao, Z.; Lu, Z.; Cai, J.; Li, X.; Jiang, X.; Zhang, D. Delamination formation, evaluation and suppression during drilling of composite laminates: A review. *Compos. Struct.* **2019**, *216*, 168–186. [CrossRef]
4. Burkov, M.; Eremin, A. Hybrid CFRP/SWCNT Composites with Enhanced Electrical Conductivity and Mechanical Properties. *J. Mater. Eng. Perform.* **2018**, *27*, 5984–5991. [CrossRef]
5. Gang, D. The effect of surface treatment of CF and graphene oxide on the mechanical properties of PI composite. *J. Thermoplast. Compos. Mater.* **2018**, *31*, 1219–1231. [CrossRef]
6. Singh, K.K.; Kumar, D. Experimental investigation and modelling of drilling on multi-wall carbon nanotube–embedded epoxy/glass fabric polymeric nanocomposites. *Proc. Inst. Mech. Eng. Part B J. Eng. Manuf.* **2018**, *232*, 1943–1959. [CrossRef]
7. Rafiee, M.A.; Yavari, F. Fullerene–epoxy nanocomposites-enhanced mechanical properties at low nanofiller loading. *J. Nanopart. Res.* **2011**, *13*, 733–737. [CrossRef]
8. Liang, J.; Zou, S.; Du, Q. Impact and fl exural properties of polypropylene composites reinforced with multi-walled carbon nanotubes. *Polym. Test.* **2018**, *70*, 434–440. [CrossRef]
9. Gantayat, S.; Rout, D.; Swain, S.K. Mechanical properties of functionalized multiwalled carbon nanotube/epoxy nanocomposites *Mater. Today Proc.* **2017**, *4*, 4061–4064. [CrossRef]
10. Watson, G.; Starost, K.; Bari, P.; Faisal, N.; Mishra, S.; Njuguna, J. Tensile and Flexural Properties of Hybrid Graphene Oxide/Epoxy Carbon Fibre Reinforced Composites. *IOP Conf. Ser. Mater. Sci. Eng.* **2017**, *195*, 012009. [CrossRef]
11. Cho, J.; Chen, J.Y.; Daniel, I.M. Mechanical enhancement of carbon fiber/epoxy composites by graphite nanoplatelet reinforcement. *Scr. Mater.* **2007**, *56*, 685–688. [CrossRef]
12. Abdullah, S.I.; Ansari, M.N.M. Mechanical properties of graphene oxide (GO)/epoxy composites. *HBRC J.* **2015**, *11*, 151–156. [CrossRef]
13. Ghosh, S.; Nitin, B.; Remanan, S.; Bhattacharjee, Y.; Ghorai, A.; Dey, T.; Das, T.K.; Das, N.C. A Multifunctional Smart Textile Derived from Merino Wool/Nylon Polymer Nanocomposites as Next Generation Microwave Absorber and Soft Touch Sensor. *ACS Appl. Mater. Interfaces* **2020**, *12*, 17988–18001. [CrossRef] [PubMed]
14. Ghosh, S.; Ganguly, S.; Das, P.; Das, T.K.; Bose, M.; Singha, N.K.; Das, A.K.; Das, N.C. Fabrication of Reduced Graphene Oxide/Silver Nanoparticles Decorated Conductive Cotton Fabric for High Performing Electromagnetic Interference Shielding and Antibacterial Application. *Fibers Polym.* **2019**, *20*, 1161–1171. [CrossRef]
15. Davim, J.P.; Reis, P. Damage and dimensional precision on milling carbon fiber-reinforced plastics using design experiments. *J. Mater. Process. Technol.* **2005**, *160*, 160–167. [CrossRef]

16. Fleischer, J.; Teti, R.; Lanza, G.; Mativenga, P.; Möhring, H.C.; Caggiano, A. Composite materials parts manufacturing. *CIRP Ann.* **2018**, *67*, 603–626. [CrossRef]
17. Hintze Wolfgang, W.; Hartmann, D.; Schütte, C. Occurrence and propagation of delamination during the machining of carbon fibre reinforced plastics (CFRPs)—An experimental study. *Compos. Sci. Technol.* **2011**, *71*, 1719–1726. [CrossRef]
18. Voss, R.; Seeholzer, L.; Kuster, F.; Wegener, K. Influence of fibre orientation, tool geometry and process parameters on surface quality in milling of CFRP. *CIRP J. Manuf. Sci. Technol.* **2017**, *18*, 75–91. [CrossRef]
19. Azmi, A.I.; Lin, R.J.T.; Bhattacharyya, D. Machinability study of glass fibre-reinforced polymer composites during end milling. *Int. J. Adv. Manuf. Technol.* **2013**, *64*, 247–261. [CrossRef]
20. Bagci, E.; Yüncüoğlu, E.U. The effects of milling strategies on forces, material removal rate, tool deflection, and surface errors for the rough machining of complex surfaces. *Stroj. Vestnik/J. Mech. Eng.* **2017**, *63*, 643–656.
21. Wang, C.; Wen, L.; Ming, W.; An, Q.; Chen, M. Experimental study on effects of fiber cutting angle in milling of high-strength unidirectional carbon fiber–reinforced polymer laminates. *Proc. Inst. Mech. Eng. Part B J. Eng. Manuf.* **2018**, *232*, 1813–1824. [CrossRef]
22. Jenarthanan, M.P.; Jeyapaul, R. Optimisation of machining parameters on milling of GFRP composites by desirability function analysis using Taguchi method. *Int. J. Eng. Sci. Technol.* **2018**, *5*, 22. [CrossRef]
23. Adak, N.C.; Chhetri, S.; Sabarad, S.; Roy, H.; Murmu, N.C.; Samanta, P.; Kuila, T. Direct observation of micro delamination in graphene oxide incorporated carbon fiber/epoxy composite via in-situ tensile test. *Compos. Sci. Technol.* **2019**, *177*, 57–65. [CrossRef]
24. Wang, C.; Li, J.; Sun, S.; Li, X.; Zhao, F.; Jiang, B.; Huang, Y. Electrophoretic deposition of graphene oxide on continuous carbon fibers for reinforcement of both tensile and interfacial strength. *Compos. Sci. Technol.* **2016**, *135*, 46–53. [CrossRef]
25. Shil, A.; Roy, S.; Balaji, P.S.; Kumar Katiyar, J.; Pramanik, S.; Kumar Sharma, A. Experimental analysis of mechanical properties of stir casted aluminium-graphene nanocomposites. *IOP Conf. Ser. Mater. Sci. Eng.* **2019**, *653*, 012021. [CrossRef]
26. Bhawal, P.; Ganguly, S.; Chaki, T.K.; Das, N.C. Synthesis and characterization of graphene oxide filled ethylene methyl acrylate hybrid nanocomposites. *RSC Adv.* **2016**, *6*, 20781–20790. [CrossRef]
27. Arun, G.K.; Sreenivas, N.; Reddy, K.B.; Krishna Reddy, K.S.; Shashi Kumar, M.E.; Pramod, R. Investigation on Mechanical Properties of Graphene Oxide reinforced GFRP. *IOP Conf. Ser. Mater. Sci. Eng.* **2018**, *310*, 012158. [CrossRef]
28. Palanikumar, K.; Karunamoorthy, L.; Karthikeyan, R. Multiple performance optimization of machining parameters on the machining of GFRP composites using carbide (K10) tool. *Mater. Manuf. Process.* **2006**, *21*, 846–852. [CrossRef]
29. Wang, F.J.; Zhang, B.Y.; Zhao, M.; Cheng, D.; Wang, Z.G. Evolution laws of fiber-matrix interface cracks in machining of carbon fiber reinforced polymer. *Int. J. Adv. Manuf. Technol.* **2019**, *101*, 963–977. [CrossRef]
30. Hossain, M.S.; Dhar, N.R. Machinability Study of a Kevlar and Glass Reinforced Polyester. *J. Mech. Eng.* **2019**, *49*, 18–26.
31. Wang, H.; Sun, J.; Zhang, D.; Guo, K.; Li, J. The effect of cutting temperature in milling of carbon fiber reinforced polymer composites. *Compos. Part A Appl. Sci. Manuf.* **2016**, *91*, 380–387. [CrossRef]
32. Palanikumar, K. *Analyzing Surface Quality in Machined Composites*; Woodhead Publishing Limited: Cambridge, UK, 2012.
33. Ha, S.J.; Kim, K.B.; Yang, J.K.; Cho, M.W. Influence of cutting temperature on carbon fiber-reinforced plastic composites in high-speed machining. *J. Mech. Sci. Technol.* **2017**, *31*, 1861–1867. [CrossRef]
34. Saravanakumar, K.; Subramanian, H.; Arumugam, V.; Dhakal, H.N. Influence of milled glass fillers on the impact and compression after impact behavior of glass/epoxy composite laminates. *Polym. Test.* **2019**, *75*, 133–141. [CrossRef]
35. Papadopoulos, A.; Gkikas, G.; Paipetis, A.S.; Barkoula, N.M. Effect of CNTs addition on the erosive wear response of epoxy resin and carbon fibre composites. *Compos. Part A Appl. Sci. Manuf.* **2016**, *84*, 299–307. [CrossRef]
36. Wang, X.; Melly, S.K.; Li, N.; Wang, G.D.; Peng, T.; Li, Y.; Zhao, Q. Di Helical milling response of glass fiber-reinforced polymer composite with carbon nanotube buckypaper interlayer. *Polym. Polym. Compos.* **2020**, *28*, 378–387.
37. Razfar, M.R.; Zadeh, M.R.Z. Optimum damage and surface roughness prediction in end milling glass fibre-reinforced plastics, using neural network and genetic algorithm. *Proc. Inst. Mech. Eng. Part B J. Eng. Manuf.* **2009**, *223*, 653–664. [CrossRef]
38. Kumar, J.; Verma, R.K. Experimental investigations and multiple criteria optimization during milling of graphene oxide (GO) doped epoxy/CFRP composites using TOPSIS-AHP hybrid Module. *FME Trans.* **2020**, *48*, 628–635. [CrossRef]
39. Kumar, J.; Verma, R.K. Delamination assessment during machining of laminated polymer nanocomposite. *Int. J. Mod. Manuf. Technol.* **2021**, *13*, 109–115. [CrossRef]

Journal of
Composites Science

MDPI

Article

Hole Quality Observation in Single-Shot Drilling of CFRP/Al7075-T6 Composite Metal Stacks Using Customized Twist Drill Design

Jebaratnam Joy Mathavan [1,2], Muhammad Hafiz Hassan [1,*], Jinyang Xu [3,*] and Gérald Franz [4]

1 School of Mechanical Engineering, Universiti Sains Malaysia, Nibong Tebal 14300, Malaysia
2 Department of Engineering Technology, Faculty of Technology, University of Jaffna, Kilinochchi Premises, Ariviyal Nagar, Kilinochchi 44000, Sri Lanka
3 State Key Laboratory of Mechanical System and Vibration, School of Mechanical Engineering, Shanghai Jiao Tong University, Shanghai 200240, China
4 Laboratoire des Technologies Innovantes, UR UPJV 3899, Avenue des Facultés, Le Bailly, 80025 Amiens, France
* Correspondence: mhafizhassan@usm.my (M.H.H.); xujinyang@sjtu.edu.cn (J.X.)

Abstract: In the modern aircraft manufacturing industry, the use of fiber metal stack-up material plays an important role. During assembly, these stack-up materials need to be drilled, and single-shot drilling is the best option to avoid misalignments. This paper discusses hole quality in terms of hole edge defects and hole integrity with respect to tool geometry. In this study, tungsten carbide (WC) twist-type drills with various geometric features were fabricated, tested, and evaluated. Twenty custom twist drill bits with primary clearance angles ranging from 6° to 8°, chisel edge angles from 30° to 45°, and point angles from 130° to 140° were fabricated. The CFRP and Al 7075-T6 were stacked up, and a feed rate of 0.05 mm/rev and spindle speed of 2600 rev/min were used for all drilling experiments. The experimental array was constructed using response surface methodology (RSM) to design the experiments. The impact of factors and their importance on hole quality were investigated using analysis of variance (ANOVA). The study demonstrates that the primary clearance angle, followed by the chisel edge angle, is the most important factor determining hole quality. As a function of tool geometry, correlation models between exit delamination and burr height were developed. The findings suggested that, within the range of parameters examined, the proposed correlation models might be utilized to predict performance measures. For drilling CFRP/AL7075-T6 stack material in a single shot, the ideal twist drill geometry was determined to be a 45° chisel edge angle, 8° primary clearance angle, and 130° point angle. For optimum drill geometry, the discrepancy between the expected and actual experiment values was 0.11% for exit delamination and 9.72% for burr height. The findings of this research elucidate the relationship between tool geometry and hole quality in single-shot drilling of composite-metal stacks, and more specifically, they may serve as a useful, practical guide for single-shot drilling of CFRP/Al7075-T6 stack for the manufacture of aircraft.

Keywords: single-shot drilling; CFRP/Al stacks; hole quality; optimization; twist drill; ANOVA

Citation: Joy Mathavan, J.; Hassan, M.H.; Xu, J.; Franz, G. Hole Quality Observation in Single-Shot Drilling of CFRP/Al7075-T6 Composite Metal Stacks Using Customized Twist Drill Design. *J. Compos. Sci.* **2022**, *6*, 378. https://doi.org/10.3390/jcs6120378

Academic Editor: Francesco Tornabene

Received: 8 October 2022
Accepted: 2 December 2022
Published: 8 December 2022

1. Introduction

Composite materials have gained prominence during the past few years as a means of reducing the weight of aircraft structures. In actuality, 52% of the Airbus A350's total structural materials and 57% of the Boeing 787's major structure are composites [1], and carbon fiber-reinforced polymer (CFRP) is the most widely utilized fiber [2]. Fibrous composites and metallic alloys, such as titanium/aluminum alloys, are widely employed in stack form in the current aerospace sector to gain enhanced mechanical properties and function for components requiring energy-saving features [3–5]. Composites have enormous application potential in a variety of modern commercial aircraft, including the Airbus A350 and Boeing 787 Dreamliner [6,7]. Although there are many fibers and metals, the superior qualities of CFRP, Al, and Ti are attractive for this application and typically, while manufacturing a stack up, a composite panel is placed on top of the metal part [8]. To produce different geometric characteristics for improved product integrity, reliability, life cycle, and secure assembly with other components, hole drilling has long been a standard procedure in the manufacturing sector [9,10]. When it comes to the machining of lightweight metals and composites, this method has been particularly important in the automotive and aerospace industries [11,12]. Given that metals are isotropic and fibers are anisotropic, single-shot stack drilling involving these two components is extremely difficult and can result in a variety of tool and hole damage. This research intends to determine appropriate tool geometry improvements in regard to hole quality.

Delamination is a main issue related to the drilling of fiber-reinforced composite materials, which tends to decrease the structural integrity of the material [13]. Delamination damage is a force-associated failure that can be classified into two types [14], namely peel-up delamination and push-out delamination [1]. Peel-up delamination is caused by tool geometry [15], whereas push-out delamination is caused by thrust force inserted by the drill point [16]. Because of the irreparable nature of delamination damage, the composite laminate has to be rejected when delamination reaches a certain extent [17]. As a consequence of the inhomogeneity of fibrous composites, measurement of the extent of delamination becomes challenging. The hardness of carbon fibers induces abrasive wear at the cutting edge of the drill, which in turn increases the thrust force during drilling, finally leading to delamination [18]. On the other hand, high drilling thrust force can also be generated by increased drill feed [19]. Conversely, research carried out on dry drilling of CFRP/Al/CFRP by ref. [20] found that an increase in feed rate resulted in a positive influence on entrance delamination. Faster chip evacuation is caused by the selection of higher axial feed, thus reducing contact time and friction [21]. Delamination initiates from the CFRP matrix laminate side of the stack-up material. Although a defect-less CFRP element is formed to a near net shape, delamination at the exit of the hole is inevitable during the drilling of rivet holes. Delamination weakens the structural consistency of the CFRP part in terms of tensile and bearing strength [22], and it may also reduce the fatigue life [23].

Burr formation is a challenging factor in the aircraft industry related to multi-material stack drilling because rough edges (commonly named burrs) on fastener holes can induce stress concentrations, which may initiate corrosion, fatigue failure, reduction in the life of the aircraft [24], injuries to workers, and can reduce the functionality of the components [25]. Although burr height is the commonly measured parameter for evaluating burrs, burr thickness causes more deburring costs than burr height [26]. Usually, exit burr height and exit burr root thickness are noticeably larger than those of entrance burrs. This is primarily because the burr formed at the entrance is caused by a tearing action, which includes a bending process followed by lateral extrusion or clean shearing, whereas the exit burr is formed by plastic deformation of the workpiece material in front of the chisel edge, without the material being cut [24]. This is because the ductility of the aluminum alloy increases due to thermal softening from the higher machining temperature at higher spindle speeds. The increase in ductility allows the material to flow easily and at this stage, as the tool exits from the hole, the aluminum material is stretched and pushed out to form a burr along the edge of the hole [27]. An increasing point angle and larger helix angle tend to reduce burr root thickness and burr height [28].

Because of the tight tolerance of the hole diameter in the assembly process of the aircraft, the difference in hole diameters between the stack-up materials during drilling is an important problem. This difference in diameter occurs because of the different properties of the stack-up materials, including their elasticity modulus, which leads to different elastic deformations that make it difficult to control the difference in hole diameters between the stack-up materials [29–32]. Even if the hole in one material of the stack is undersized (hole diameter < tool diameter) or oversized (hole diameter > tool diameter), a reparation technique needs to be applied that usually adds extra costs and time to the assembly procedure. Soo et al. [31] used two types of twist drill designs, namely flat point and double cone, to drill CFRP/Al stack material in a single shot, revealing that the double cone drill bit design helped control the difference in hole diameter between both stack materials due to thin chip formation and easy evacuation of chips through the drill flute during the drilling process. Benezech et al. [33] mentioned the importance of the axial rake angle and the included angle on drilled hole quality during CFRP-Al stack drilling. These authors kept a constant axial rake angle throughout the length of the cutting edge, since it is advantageous for good quality drilling. The combination of 135° point angle and 30° rake angle gave the optimum twist drill geometry to attain high quality holes in stack-up material. Kuo et al. [34] investigated the number of margins in the twist drill bit design that significantly influenced the diameter of the hole, regardless of the feed rate and drilling technique. Triple-margin drills gave less vibration and larger contact with the machined surface, yielding smaller hole diameter variation. The higher ductility of the aluminum alloy resulted in a change to long, twisted helical chips, but they were tightly folded as the drill progressed into the stack. Accordingly, oversized holes can simply arise when the drilling operation is executed in dry conditions.

To date, the influence of the geometric parameters of twist drills on hole quality have not been reported by previous researchers. This research work highlights the single effects of the geometric parameters of the drill bit on hole quality. This paper is written in a format that provides a brief introduction to the relationship between drilling parameters and hole quality in Section 1. Section 2 describes the materials and methods used in this work Section 3 contains the results and discussion, and finally, Section 4 delivers the conclusion of this research work.

2. Materials and Methods

2.1. Worpiece Materials

CFRP and 7075-T6 aluminum alloy (Al7075-T6) were the stack materials used in this study. The CFRP composite specimen had a total laminate thickness of 3.25 mm and was made up of 26 unidirectional plies, each 0.125 mm in thickness. Hexcel Composite Company's carbon/epoxy prepreg was used to make the 26 unidirectional plies with a stacking sequence of $[45/135/90_2/0/90/0/90/0/135/45_2/135]_S$. The CFRP laminate was then covered with a 0.08 mm thin layer of glass/epoxy woven fiber at the top and bottom to prevent delamination at both the entrance and exit of the hole during drilling. As a result, the final thickness of the entire composite panel, including the paint application, was 3.587 mm. The CFRP was compressed using a vacuum pump and controlled atmospheric conditions during the curing process. The autoclave was equipped with a prepared mold to keep the laminate. The temperature was raised to 180 °C during the curing cycle at a rate of 3 °C/min and maintained for 120 min. The temperature was then gradually brought back to normal temperature. The entire cycle was carried out in an autoclave at a pressure of 700 kPa and the laminate was packed in a vacuum bag that was depressurized to 70 kPa. Because of the curing recipe's application, the nominal fiber volume was 60% The mechanical and physical characteristics of the stack materials employed in this work are compiled in Table 1.

Table 1. Mechanical properties of CFRP and Al7075-T6.

Properties	Tensile Strength [MPa]	Elasticity Module [GPa]	Elongation [%]	Flexural Strength [MPa]	Density [g/cm^3]	Thickness [mm]
CFRP	2723	164	1.62	1500	1.601	3.587
Al7075-T6	558	71.7	13	-	2.597	3.317

2.2. Cutting Tool Fabrication

The drill bit type was a combination of drill and countersink. The drill's diameter was 4.826 mm and the diameter of the countersink was 10 mm. Due to its excellent resistance to wear while drilling abrasive materials like CFRP, a sintered rod of tungsten carbide (WC) was chosen as the drill bit material. The tungsten carbide rod was made up of 93.36 wt% WC and 6.64 wt% Cobalt (Co). Since tungsten carbide has a Vickers hardness value of 1625 HV and density of 14.35 g/cm^3, both of which are much higher than those of the workpiece, it was selected as the drill bit material. Helitronic Tool Studio version 1.9.216.0 software (Walter Maschinenbau GmbH, Garbsen, Germany) was used to design the drills with special custom drill geometry. A particular wheel must perform numerous consecutive operations while grinding with a cutting tool. These operations include pointing, gashing, and clearing. In the program, a chisel edge angle of 30° to 45° was set for the gashing process and a primary clearance angle of 6° to 8° was selected for the clearance phase. The point angle was finally established from 130° to 140° during the pointing phase. Figure 1a–c demonstrates the manufacturing procedure and the wheel type used to modify the twist drill design using a CNC grinding machine (Walter Maschinenbau GmbH, Garbsen, Germany).

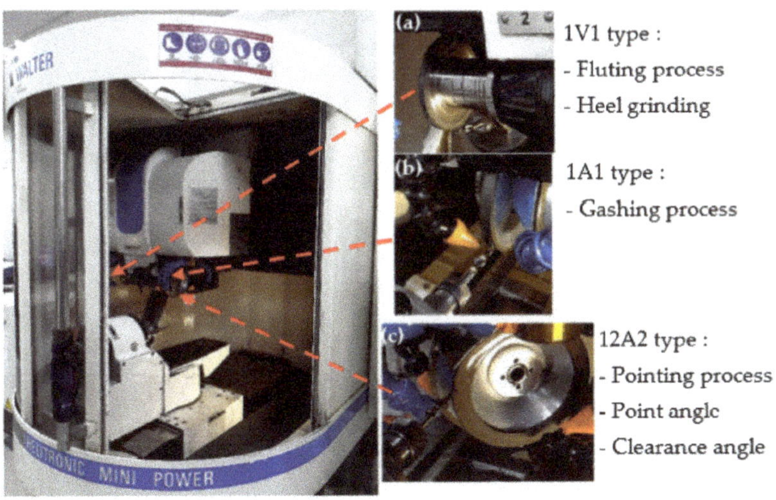

Figure 1. Location of grinding wheel for the tool fabrication: (**a**) fluting wheel, (**b**) gashing wheel, (**c**) clearance/point angle wheel.

2.3. Drilling Process

Using a computer numerical control (CNC) machine (Fanuc Robodrill T21iFLb), which has a variable spindle speed up to 10,000 rev/min and spindle drive motor of 3.7 kW at a continuous rating, the drilling of the stack material was carried out. For a regular rate, the feed rate can range from 1 to 30 mm/min, and for a high transverse rate, the feed rate can range from 48 m/min (x, y, and z axes). Drilling was performed in a single shot, starting at the CFRP panel and moving to the Al7075-T6 panel. During drilling, the stack panels were slotted into the fixture and clamped. To evaluate the major impact of the customized twist drill geometry, a feed rate of 0.05 mm/rev and spindle speed of 2600 rev/min were chosen for all runs in this study. In this experiment, dry drilling conditions were employed to simulate the drilling process that actually occurs during panel manufacturing. The angles of the standard twist drills, with variations in the three aforementioned angles, are summarized in Table 2. Design of experiment (DOE) was used to design the experimental process. DOE is a popular technique for constructing the number of experiments needed to establish the statistical validity of the relationship between the input and output of independent variables. Twenty trials were administered in accordance with ref. [35].

Table 2. Experimental factors at different levels of chisel edge angle (A), primary clearance angle (B), and point angle (C).

Level	Chisel Edge Angle [°]	Primary Clearance Angle [°]	Point Angle [°]	Spindle Speed [rev/min]	Feed Rate, [mm/rev]
Minimum	30	6	130		
Midpoint	37.5	7	135	2600	0.05
Maximum	45	8	140		

2.4. Hole Edge Defect Measurement

The quality of a drilled hole in the aircraft industry can be defined based on the hole edge defects and hole integrity. Delamination in the CFRP phase and burr formation on the aluminum phase are the major hole edge defects. Poor hole edges can contribute to stress formation and rivet joint damage during mounting.

2.4.1. Exit Delamination

Delamination in this research was evaluated at the exit side of the CFRP laminate. Figure 2a shows the sample CFRP panel condition at the exit side after the drilling process. The laminate was assessed using an Alicona InfiniteFocus optical microscope at 20× magnification to observe the delamination at the exit side of the CFRP panel in detail (Figure 2b). To measure the value of the delamination of the CFRP laminate at the end of the drilling process, a delamination factor was introduced. To make sure that the delamination was within the specification limits according to OEM standards, the delamination in the bore of the drilled hole at the exit hole face of the CFRP section must be less than 2 mm per side for laminates with a thickness of less than 5 mm. The images from the Alicona InfiniteFocus optical microscope were investigated using ImageJ software in order to determine the area of nominal value and the damaged area. The delamination factor, F_{d-exit}, was calculated based on the ratio of the damaged area to the nominal area, as shown in Equation (1).

$$F_{d-exit} = \frac{A_{max}}{A_{nom}} \tag{1}$$

where A_{nom} is the nominal area of the drilled hole and A_{max} is the damaged area of the composite laminate after the drilling process.

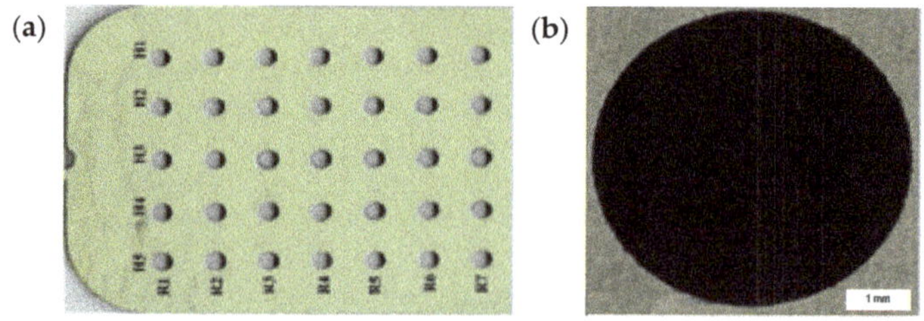

Figure 2. Observation of (**a**) exit (**b**) close-up view of delamination at the exit using an Alicona InfinateFocus optical microscope.

2.4.2. Burr Height

Minimizing burr formation is a vital criterion in drilling. The current study focused only on exit side burrs, since these generally lead to further processes such as dismantling, deburring, and reassembly of the stack, while entry side burrs are not significant because of the compaction force from the CFRP laminate. The primary factors that affect burr formation are cutting parameters, tool geometry, and workpiece materials. The smallest burrs at the hole edge were observed by ref. [36] when the drilling feed rate was increased. The tendency for burr formation may increase if the material has moderate ductility, since the material tends to elongate because of the produced heat during the machining process.

Exit burr formation in this study was evaluated using an Alicona Infinite Focus optical microscope with a magnification of 20×, as shown in Figure 3. The optical measurement system was a non-contact type that accomplished the task without creating any surface damage. The detailed maximum burr formation measurement is shown in Figure 4. The maximum burr height was identified from the 3D diagram obtained from the Alicona InfiniteFocus optical microscope, and the highest burr point is marked by a red line, as shown in Figure 4c.

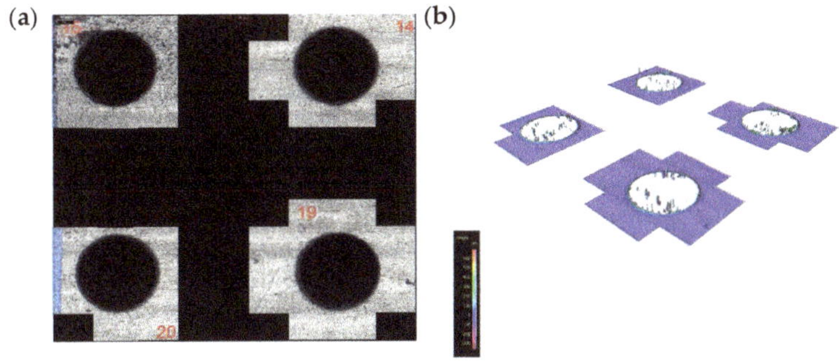

Figure 3. Type of burr formation observed under the Alicona Infinite Focus optical microscope: (**a**) uniform burr formation, (**b**) rolled-back burr formation.

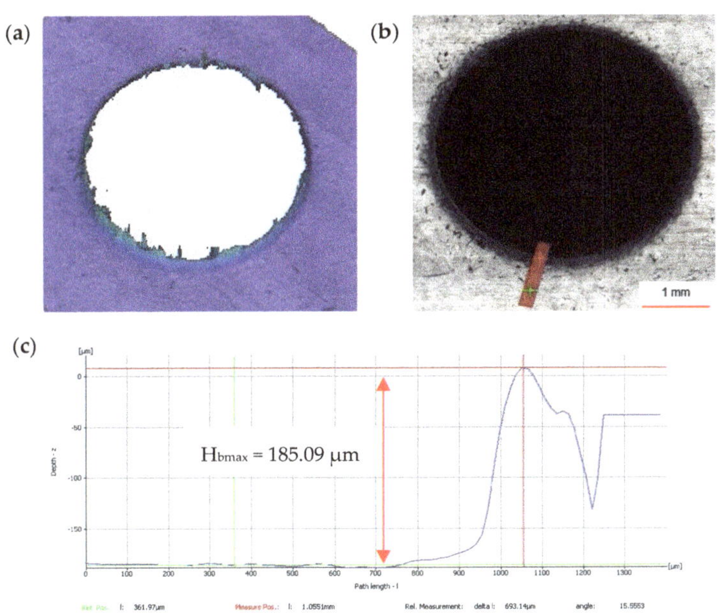

Figure 4. Measurement process of maximum burr formation: (**a**) 3D observation of burr formation, (**b**) maximum burr formation identification, and (**c**) maximum burr formation measurement.

2.5. Hole Integrity

Hole integrity is defined based on hole circularity or hole roundness and the difference in hole diameters between materials. Hole circularity defects and large differences in hole diameters between the stacking materials will also interrupt the assembly process, which in succession will increase the quantity of scrap panels. These parameters are very important and must be scrutinized frequently according to the requirements of the customer.

2.5.1. Hole Diameter Error

Ensuring the minimum difference in diameter between holes in CFRP and Al, which have different material properties, is one of the main tasks while drilling a stack-up material The definition of hole diameter error is the difference of measured hole diameter to the nominal diameter. In the case of the stack-up material in this research work, hole diameter error was defined as the difference between the measured holes of CFRP and Al (Al7075-T6) A Crysta-Plus M443 coordinate measuring machine (CMM) with a probe size of 2 mm and accuracy in measuring error of 3.0 + 4 L/1000 μm was used in this research work to measure the hole diameter errors for both CFRP and Al7075-T6. It can function in the x-axis, y-axis, and z-axis, commonly termed the three orthogonal axes, in a three-dimensional coordinate system. After the probe touched the surface, the point positions acted as the input and the data were transmitted to MCOSMOS v3 software.

Then, the software employed the x-axis, y-axis, and z-axis coordinates of every discrete point to find the mean diameter of the hole. The workpiece sample was raised up by a block, clamped, and placed in a position where it could be reached for measurements of the x-axis, y-axis, and z-axis. The coordinate system was arranged in a way to choose a datum point as a reference. The measurement of hole diameters started with the CFRP panel. For laminates ranging in thickness from 3 to 10 mm, the probe position should be in the center of the laminate, according to OEM standards. The probe in this research work was moved towards the center of the drilled hole and then downwards into half of the hole depth of the laminate thickness, as shown in Figure 5a. The positions of 0°, 90°, 180°, and 270° points were obtained as four reference points during the measurement to confirm the consistency of this procedure, as shown in Figure 5b. A circle appeared on the screen and the diameter was recorded. These steps were repeated for Al7075-T6 until the diameters of all the holes of the sample were measured, as shown in Figure 6. The following Equations (2)–(4) were used to calculate the hole diameter error for each panel and also between the laminates.

$$\varepsilon_{cfrp} = d_m - d_{nom} \tag{2}$$

$$\varepsilon_{al\ 7075} = d_m - d_{nom} \tag{3}$$

$$\varepsilon_{cfrp/al\ 7075} = d_{cfrp} - d_{al\ 7075} \tag{4}$$

where ε_{cfrp} is the error for CFRP panel; $\varepsilon_{al\ 7075}$ is the error for Al7075-T6 panel; $\varepsilon_{cfrp/al\ 7075}$ is the difference in diameter between stack laminates; d_m is the measured diameter; d_{nom} is the nominal diameter; d_{cfrp} is the measured diameter for CFRP; and $d_{al\ 7075}$ is the measured diameter for Al7075-T6 panel.

2.5.2. Hole Circularity

Hole circularity was measured using the Crysta-Plus M443 CMM in the same way as the hole diameter was measured. The only difference between the measurement of hole diameter and that of hole circularity was the number of points obtained for consideration. At least 40 points should be measured to obtain the least square diameter and circularity of a hole at a given depth [37]. A sample measurement and the point distribution to obtain hole circularity are shown in Figure 7. The hole circularity values of CFRP and Al7075-T6 were individually obtained from the information given in Figure 6 after all 40 points were touched.

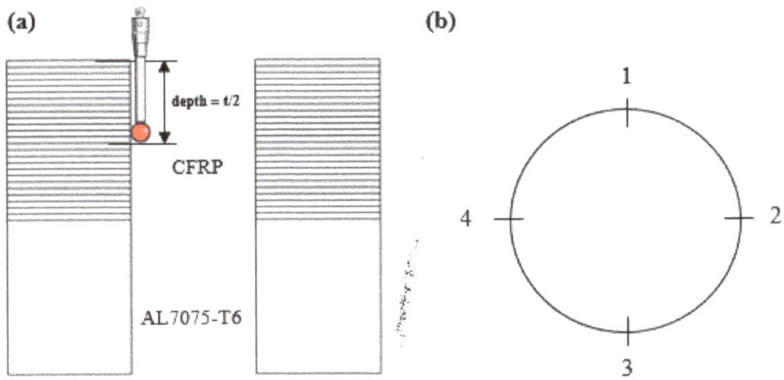

Figure 5. (**a**) Position of probe during measurement of each laminate CFRP and Al7075-T6 (**b**) point of contact for hole diameter measurement and hole circularity measurement.

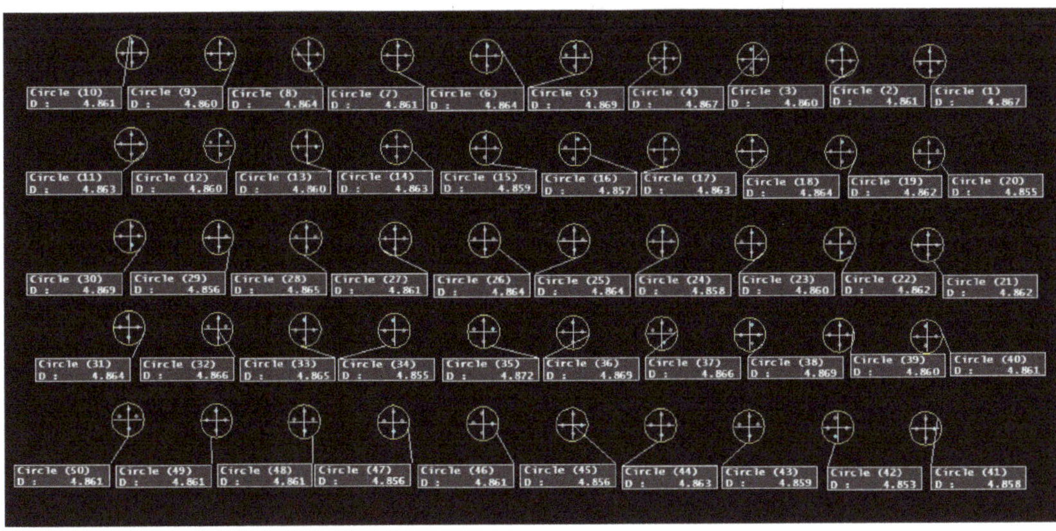

Figure 6. Example of hole diameter measurement for Al7075−T6 panel.

2.6. Response Surface Methodology (RSM)

RSM [35] has several advantages over Taguchi's method and is a crucial tool for optimizing a product or process and solving resilient design problems [38]. RSM was used to modify the twist drill geometry for single-shot drilling of the stack-up material. In this optimization study, interactions between the input variables were evaluated. The most common of all second-order designs, the central composite design (CCD), was used in the RSM design. The CCD comprises a full factorial design (2k) with 2k of axial or star points and center points, where k is the number of factors [39]. Between −1 and +1, there are three levels of variables. Table 3 lists the parameters that were selected for this investigation along with their coding levels. The formula CCD = 2^k + 2k + 6, was used to produce the number of experiment runs, where k was the number of components with replications at the design center. A quadratic model was applied to the optimization to fit and estimate the minimal point. The mathematical model for each answer was created using these data points, as illustrated in Equation (5) [40,41].

$$Y = \beta_0 + \sum_i^k \beta_i X_i + \sum_{i=1}^k \beta_{ii} X_i^2 + \sum_{i=1}^k \sum_{j=1}^k \beta_{ij} X_i X_j \quad (5)$$

where, Y is the predicted response; X_i and X_j are the input variables; β_0 is an offset term; β_i, β_{ii}, and β_{ij} are the interaction coefficients of linear, quadratic, and second-order terms.

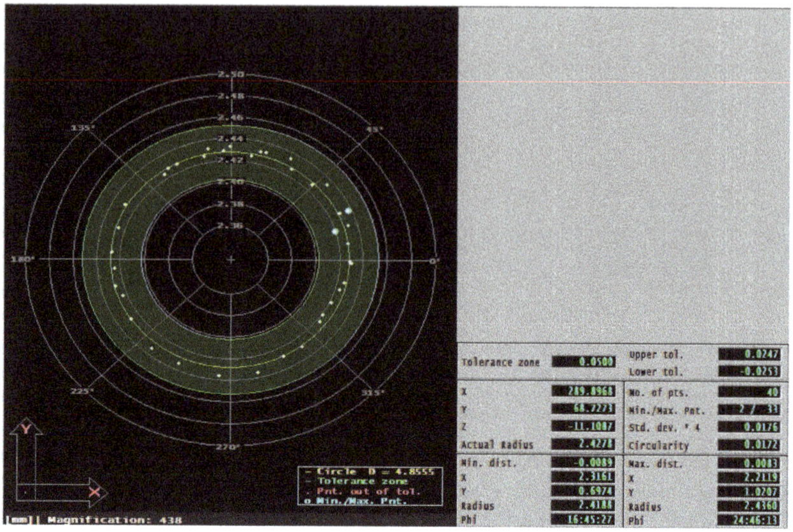

Figure 7. Example of hole circularity measurement of CFRP panel.

Table 3. The parameters selected for RSM investigation along with their coding levels.

Input Variables	Lower Level (−1)	Coded Level (0)	Higher Level (+1)
Chisel Edge Angle [A°]	30	37.5	45
Primary Clearance Angle [B°]	6	7	8
Point Angle [C°]	130	135	140

Design Expert 14 software was used to choose the regression models for the results based on the highest-order polynomials, significant additional terms, and lack of aliased models. The regression models were created in terms of coded and real components, with the best fitting of the quadratic equation or other transformation for all relevant model variables taken into consideration at values of p-values less than 0.05 [42]. Through the perturbation plot, the responsiveness of independent variables was determined. To determine the relationships between parameters, two significant components were chosen to create the 3D response surface. The intended aim, whether to minimize or enhance the output, was determined in accordance with the level's range, following the optimization process. The software then produced each factor's optimal value and reaction. The outcome of the experiment was then compared to the outcome of the regression models, which were constructed using the best value possible for each factor.

3. Results and Discussion

3.1. Exit Delamination Analysis

Figure 8a,b shows the minimum and maximum delamination, respectively, found at the CFRP panel's exit hole. Although the point angle of the drill bit was increased from 130° to 140°, the delamination for the entire run was within the permitted limit. This amply demonstrates that, during any of the tests, there was never a clear indication of delamination at the CFRP exit hole when the point angle was increased within the range. In a similar manner, Senthil Kumar et al. [43] used 118° and 130° of point angle drills to examine the effects of point angle on tool performance when drilling composite/Ti stack. It was determined that the higher point angle (130°) drills outperformed those with the lower point angle (118°), based on tool wear and chip evacuation analysis. Geng et al. [1] demonstrated that the drilling thrust force exceeds the critical thrust force when exit delamination occurs. Since the composite panel's critical thrust force was not reached in this research, the entire run was conducted below it.

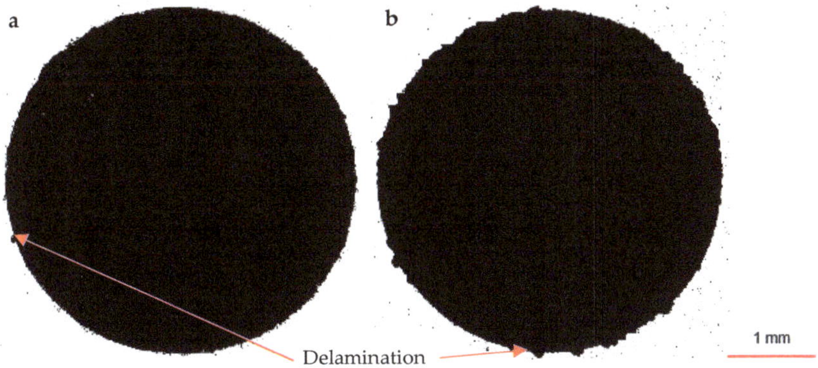

Figure 8. (**a**) Minimum and (**b**) Maximum delamination at the exit hole of the CFRP panel for all runs.

Figure 9 shows the detailed exit delamination factor $F_{d\text{-exit}}$ values for the entire run. The values acquired for each trial appeared to be nearly identical on the graph. The typical $F_{d\text{-exit}}$ value ranged from 1.0038 to 1.0196. Since every value was below the tolerance level advised by aerospace manufacturers in accordance with OEM standards, it is evident that the range of the drill geometry in this study had no impact on the $F_{d\text{-exit}}$ value. However, ref. [44] reported that twist drills are less efficient than core drills since the thrust force is not much focused on the middle of the drill bit and cutting edges, but usually distributed over the periphery of the bit.

3.1.1. Regression Model and ANOVA

To obtain the lowest residuals between the anticipated and actual values for delamination, the regression model for the response was enhanced using a quadratic model. The final empirical model for the actual causes of delamination at the CFRP panel's exit hole (Y_1) is shown in Equation (6).

$$Y_1 = 3.37924 - 2.34867e^{-3}A - 0.41360B - 0.011186C + 1.53654e^{-3}BC + 0.014163B^2 \quad (6)$$

For (Y_1), the F-value in the ANOVA analysis was 8.66 and the probability value (p-value) was less than 0.05, as shown in Table 4. Furthermore, the p-value of 0.9214 indicated that the lack of fit was related to pure error and was not significant. The model's significant value and the lack of fit's non-significant value supported the validity of the log-transformed model. The point angle was insignificant, despite the fact that the percentage of contribution (PC) for each model term, A, B, BC, and B^2, had a considerable impact on exit delamination, with values of 43.7%, 10.5%, 8.4%, and 20.4%, respectively. The values of the R^2, adjusted R^2 (Adj R^2), and predicted R^2 (Pred R^2) coefficients were used to assess the model's goodness of fit. The Adj R^2 value of 0.7053 and Pred R^2 value of 0.6885 were in reasonable accord as the discrepancy was less than 0.2.

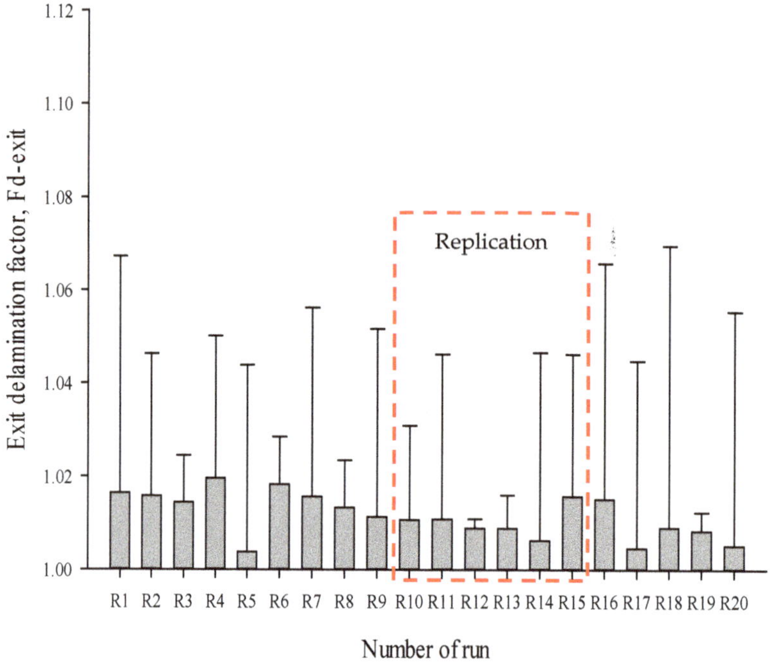

	R1	R2	R3	R4	R5	R6	R7	R8	R9	R10
Fd-exit	1.0164	1.0158	1.0144	1.0196	1.0038	1.0183	1.0156	1.0133	1.0113	1.0107
	R11	**R12**	**R13**	**R14**	**R15**	**R16**	**R17**	**R18**	**R19**	**R20**
Fd-exit	1.0108	1.0089	1.0089	1.0063	1.0157	1.0151	1.0046	1.0090	1.0083	1.0052

Figure 9. Exit delamination factor values of CFRP for all runs.

Table 4. Pooled ANOVA of model for exit delamination of CFRP panel.

Source	Sum of Squares	df	Mean Square	F Value	*p*-Value	PC (%)	
Model (Y_1)	0.0032506	5	0.0004911	8.66	0.0015		Significant
Chisel edge angle (A)	0.001695	1	0.001695	29.87	0.0002	43.7%	
Primary clearance angle (B)	0.0004053	1	0.0004053	7.15	0.0217	10.5%	
Point angle (C)	3.393×10^{-5}	1	3.393×10^{-5}	0.6	0.4556	0.9%	
BC	0.0003246	1	0.0003246	5.72	0.0357	8.4%	
B^2	0.0007918	1	0.0007918	13.96	0.0033	20.4%	
Residual	0.000624	11	5.673×10^{-5}			16.1%	
Lack of Fit	0.0001586	6	2.643×10^{-5}	0.28	0.9214		not significant
Pure Error	0.0004654	5	9.308×10^{-5}				
Cor Total	0.0038746	16					
Std. Dev.	7.53×10^{-3}		R^2		0.7974		
Mean	1.04		Adj R^2		0.7053		
C.V. %	0.73		Pred R^2		0.6885		
PRESS	9.59×10^{-4}		Adeq Precision		10.446		

A signal-to-noise ratio greater than 4 is preferred when measuring signal-to-noise with adequate precision [45]. Since a strong signal was indicated by the ratio (Adeq Precision) of 10.446, this model was utilized to navigate the design space. According to ref. [46], R^2 should be at least 0.80 for a model to fit the data well. The correlation coefficient (R^2) and adjusted coefficient (Adj. R^2) values in this instance were 0.7974 and 0.7053, respectively, demonstrating the significance of the fit of the RSM model and its potential for response prediction.

When the actual value gained through experimentation was compared with the predictions of the model, as shown in Figure 10a, it can be observed that the points were evenly split by a 45-degree line, which proved the fit of the model. This established the reliability of the regression model modifications for predicting exit delamination. The graph demonstrates that, when compared to the expected value predicted by the empirical model, the response of the experimental data was mostly contained within the range of allowable deviations, as shown in Figure 10b. When drilling stack-up material, the CFRP delamination could be estimated using the regression model created here. The standard error estimation (SEE) result was 0.0060795.

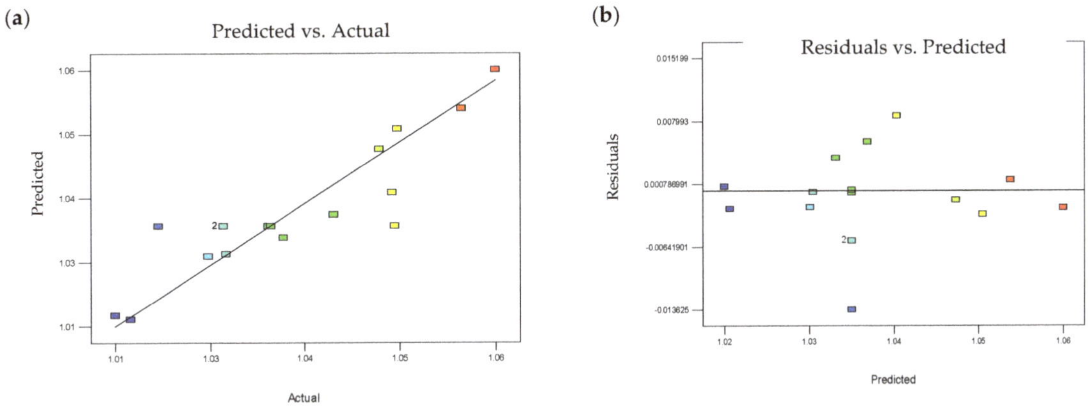

Figure 10. CFRP exit delamination analysis for (**a**) actual and predicted plot and (**b**) predicted and residual plot.

3.1.2. Effect of Geometric Parameters on Exit Delamination

For exit delamination of the CFRP (Y_1) panel, the perturbation plot shown in Figure 11a was used to determine the sensitivity of each factor. The exit delamination was significantly affected by the chisel edge angle (A). The exit delamination of the CFRP was decreased by increasing the chisel edge angle. With these drill geometries, a lower exit delamination was consequently produced. In this parameter analysis, the primary clearance angle (B) had a significantly greater effect than the point angle (C) on the exit delamination (Y_1) of the CFRP. According to the quadratic model that was fitted, a curvilinear profile was observed, as shown in Figure 11b. By maintaining the third parameter i.e., chisel edge angle constant at the middle level (37.5°), the graph indicated delamination with regard to two alternative parameters. When the point angle was set at 130° and the primary clearance angle was set at 8°, exit delamination was decreased.

3.2. Burr Height Analysis

Exit burr formation was examined using an Alicona optical microscope and burr formation was relatively uniform across the circles of the holes. The development of burrs was likely due to the accumulation of heat from the CFRP panel, which enabled the extrusion of softened Al7075-T6 at the tool margin area. The twist drill's optimal drill geometry for drilling a stack material in a single shot is one that produces the least amount of burrs because adding a second process to remove the formed burrs would raise the cost of the manufacturing process. For burrs formed at the exit of the drilled hole, the deburring process can account for approximately 30% of the total manufacturing cost and can occupy 40% of the total machining time [47]. According to Sakurai et al. [28], a large point angle ensured maximum lip movement as soon as possible to prevent work hardening, which caused thinner burrs because of a shift in chip flow direction. Low feed rates (0.05 mm/rev) are required to guarantee the least amount of thrust force in order to reduce burr development [48].

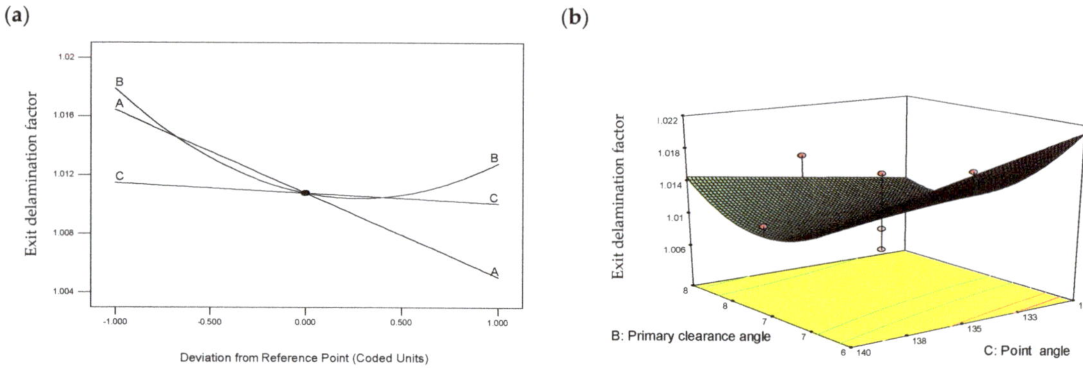

Figure 11. (**a**) Perturbation plot and (**b**) 3D Response surface for exit delamination of CFRP. A, chisel edge angle; B, primary clearance angle; C, point angle.

Figure 12 displays the measurements of the maximum and minimum exit burr formation for the typical drilled hole in Al7075-T6. The average H_{bmax} value was minimal, measuring between 40.2 and 271.2 µm, as shown in Figure 13. A lower H_{bmax} value was found in R16 with a 45° chisel edge angle, 6° primary clearance angle, and 130° point angle, whereas R3 yielded a higher H_{bmax} value with a 30° chisel edge angle, 7° primary clearance angle, and 135° point angle [35]. These results are in line with ref. [49] in which the burr height ranged from 133.62 to 211.45 µm when they used a 130° point angle and from 1036.25 to 2066.85 µm when they used a tool with a 110° point angle. Further, these authors mentioned that the drill with a 130° point angle produced a uniform burr type and a 110° point angle produced transient and crown burrs during single-shot drilling of CFRP/Al7075-T6 material [49].

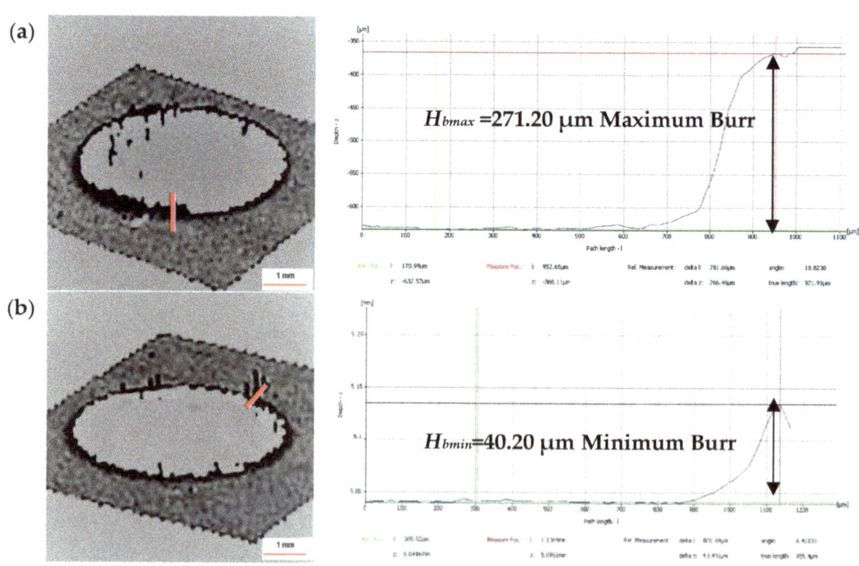

Figure 12. (**a**) Maximum and (**b**) minimum burr formation measurements.

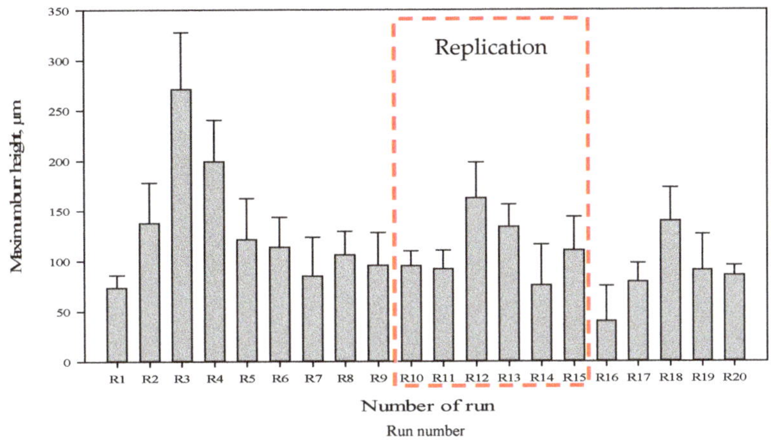

	R1	R2	R3	R4	R5	R6	R7	R8	R9	R10
H_{bmax} (µm)	73.54	137.7	271.2	199.46	121.4	113.7	84.9	106.18	95.5	95.1
	R11	**R12**	**R13**	**R14**	**R15**	**R16**	**R17**	**R18**	**R19**	**R20**
H_{bmax} (µm)	91.90	163.0	133.8	75.90	110.5	40.2	79.2	139.5	90.8	85.7

Figure 13. Burr formation data during all 20 runs.

When the chisel edge angle was reduced to 30°, the H_{bmax} value rose, resulting in a significant rolled-up phenomenon at the Al707-T6 panel's exit. This is because there was less space for the chip to flow during evacuation when the bit first contacted the material during the cutting operation, because the chisel edge angle of 30° was less than 45°, as shown in Figure 14. The ineffective chip flow increased the shear and decreased the cutting efficiency during the drilling process. The cutting heat also makes the material more ductile and uses more energy [32]. As a result, burrs along the hole's edge are easily produced. The replicated tools yielded consistent results, as shown in Figure 13, proving that they were properly manufactured.

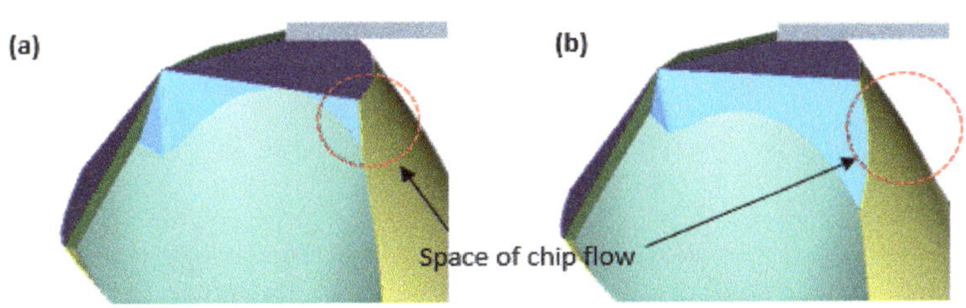

Figure 14. Space for chip flow showing two geometries with same primary clearance angle and point angle where (**a**) drill with 30° of chisel edge angle (**b**) drill with 45° of chisel edge angle.

3.2.1. Regression Model and ANOVA

To obtain the lowest residuals between the anticipated and actual values for H_{bmax}, the regression model for the response was enhanced by log transformation. The final empirical model for the actual causes of burr formation at the Al7075-T6 panel's exit (Y_2) was

$$Y_2 = -25.99962 - 0.22321A + 5.9706B + 0.1564C - 0.02019BC + 2.747e^{-3}A^2 - 0.2298B^2 \quad (7)$$

The F-value for H_{bmax} in the ANOVA analysis was 9.718 and the p-value was lower than 0.05, as shown in Table 5. Furthermore, the p-value of 0.683 indicated that the lack of fit was related to pure error and was not significant. The model's significant value and the lack of fit's non-significant value supported the validity of the log-transformed model. As indicated in Table 5, the factors that significantly influenced the results had a confidence level above 95% and a p-value lower than 0.05. The second term, B, was insignificant, despite the fact that the p-values for the other model terms (A, C, BC, A^2, and B^2) had a considerable impact on H_{bmax}. The adjusted R^2 value of 0.733 and the predicted R^2 value of 0.602 were in reasonable accord as the discrepancy was less than 0.2. Since a ratio greater than 4 is preferred when measuring the signal-to-noise ratio [45], as mentioned in Section 3.1.1, and a strong signal of 12.85 was obtained here, this model was utilized to navigate the design space. The correlation coefficient (R^2) and adjusted coefficient (Adj R^2) values in this instance were 0.818 and 0.733, respectively, demonstrating the significance of the fit of the RSM model and its potential for response prediction.

Table 5. Pooled ANOVA of model for maximum burr formation at the exit of Al7075-T6 panel.

Source	Sum of Squares	df	Mean Square	F Value	p-Value Prob > F	PC (%)	
Model (Y_2)	0.55648	6	0.08024	9.71841	0.0004		significant
Chisel edge angle (A)	0.16492	1	0.16492	19.97565	0.0006	24.8%	
Primary clearance angle (B)	0.0077	1	0.0077	0.93271	0.3518	1.2%	
Point angle (C)	0.0569	1	0.0569	6.89196	0.021	8.6%	
BC	0.08154	1	0.08154	9.87583	0.0078	12.3%	
A^2	0.07645	1	0.07645	9.25995	0.0094	11.5%	
B^2	0.16897	1	0.16897	20.46549	0.0006	25.5%	
Residual	0.10733	13	0.00826			16.2%	
Lack of Fit	0.05707	8	0.00713	0.70959	0.6832		not significant
Pure Error	0.05026	5	0.01005				
Cor Total	0.66381	19					
Std. Dev.	0.09086		R^2		0.817699		
Mean	2.02779		Adj R^2		0.733559		
C.V. %	4.48093		Pred R^2		0.601704		
PRESS	0.2345		Adeq Precision		12.8502		

When the actual value gained through experimentation was compared with the predictions of the model, as shown in Figure 15a, it can be observed that the points were evenly split by a 45-degree line, which proved the model fit. Figure 15b demonstrates that, when compared to the expected value predicted by the empirical model, the response of the experimental data was mostly contained within the range of allowable deviations. When drilling stack-up material, the Al7075-T6 burr formation could be estimated using the regression model that was created here. The standard error estimation (SEE) result for H_{bmax} $\log_{10}$ was 0.0732, according to Figure 15b. For example, the actual number fell between 1.9068 and 2.0532, and the anticipated value was 1.98. For a dataset with a normal linear relationship, the RSM model can be used to estimate the value if two-thirds of the residual data points (Figure 15b) are within SEE i.e., above or below the least squares line [50].

3.2.2. Effect of Geometric Parameters on Burr Height Formation

For H_{bmax} of the Al7075-T6 panel (Y_2), the perturbation plot shown in Figure 16a was used to determine the sensitivity of each factor. The variables had a significant impact on the specific responses in the drilling of stack-up material. The H_{bmax} value was significantly affected by the chisel edge angle (A). The H_{bmax} value at the exit of Al7075-T6 was decreased by increasing the chisel edge angle. An extreme chisel edge angle made clearance possible and made the shearing of materials by the cutting edges more effective (Figure 14). With these drill geometries, less burr formation was consequently produced. In this parameter analysis, the primary clearance angle (B) and point angle (C) had a moderate impact on the response of (Y_2).

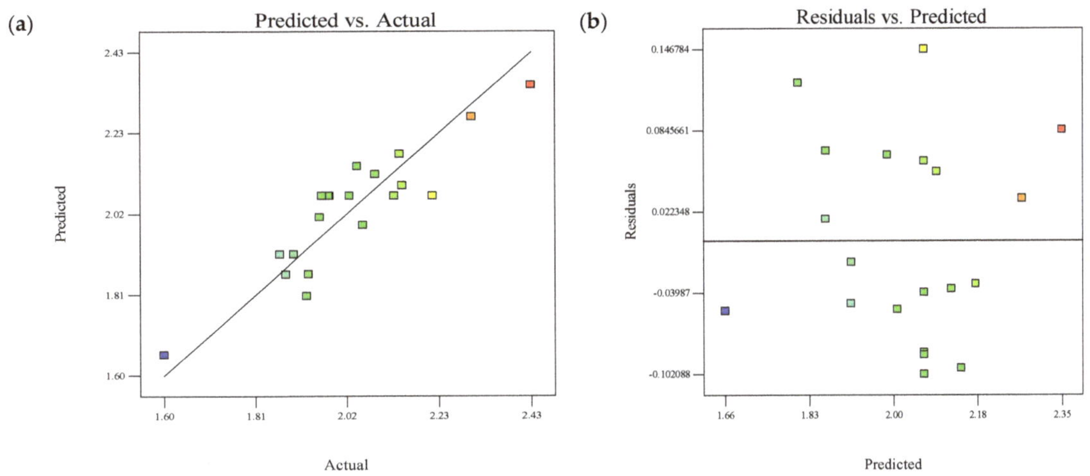

Figure 15. Analysis of maximum burr formation on Al7075−T6 for (**a**) actual and predicted plot and (**b**) predicted and residual plot.

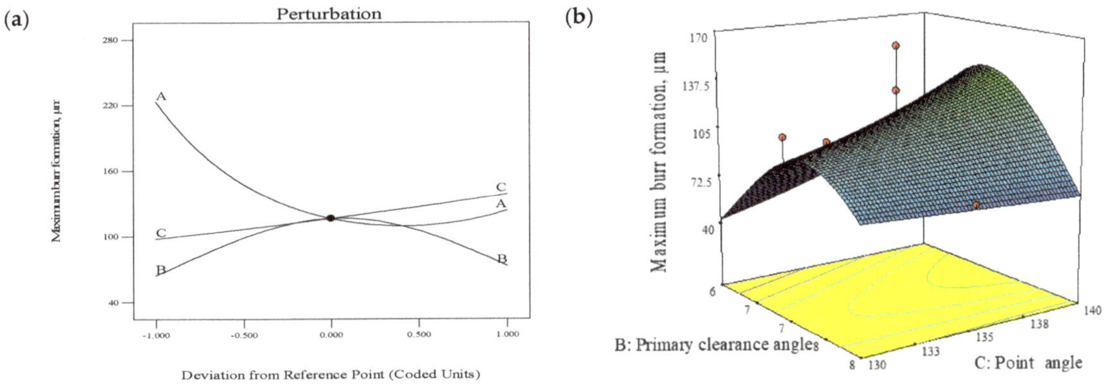

Figure 16. (**a**) Perturbation plot and (**b**) 3D response surface for exit burr height at Al7075−T6. A, chisel edge angle; B, primary clearance angle; C, point angle.

Figure 16b displays the 3D surface graphs for burr development of Al7075-T6. According to the quadratic model that was fitted, the results had a curvilinear profile. By maintaining the third parameter i.e., chisel edge angle constant at the middle level (37.5°), the graph indicated the H_{bmax} with regard to two alternative parameters. When minimum point angle and primary clearance angle were set, i.e., primary clearance angle of 6° and point angle of 130°, the H_{bmax} was decreased.

3.3. Multiple Response Optimization

This section presents the target response optimization based on the developed regression function of each response connected with the cutter geometry. Optimizing the target response facilitates achieving a set of ideal target response conditions. It can maintain all the desired response ranges or at least optimize all the desired responses. The target response technique seeks to improve quality, cost, and time, while increasing product efficiency. In this study, the target response was optimized using two techniques: an overlay plot and the desirability function. The objectives of this optimization procedure were to create a cutter with the least amount of thrust force and burr development. The goal and limitations for the variables to simultaneously attain many desired goals are tabulated in Table 6.

Table 6. Goals and constraints for the factors and responses.

Contraints			
Factor/Response	**Goal**	**Lower Limit**	**Upper Limit**
Chisel edge angle (A)	Within range	30°	45°
Primary clearance angle (B)	Within range	6°	8°
Point angle (C)	Within range	130°	140°
Burr Height (H_{bmax})	Minimize	40.2 μm	271.2 μm
Delamination ($F_{d\text{-}exit}$)	Minimize	1.0046	1.0196

Based on the goals, one solution was proposed, as tabulated in Figure 17. A desirability level closer to 1 indicates that the goals are not easy to reach. In other words, a higher desirability index represents the closest response to the target or ideal values. As shown in Figure 17, the proposed solution gave a desirability index of 0.773.

Solutions

Number	Chisel edge ar	Primary cleara	Point angle	Burr Height	Delamination	Desirability	
1	45.0	8	130	82.2307	1.00528	0.773	Selected
2	44.9	8	130	81.9794	1.00532	0.773	
3	44.8	8	130	81.4022	1.00542	0.772	
4	45.0	8	130	82.0924	1.00533	0.772	
5	44.6	8	130	80.5027	1.00559	0.771	
6	44.5	8	130	80.245	1.00564	0.771	

Figure 17. Proposed solution report for the optimization tool geometry process.

In this this experiment, the optimal cutter geometry (45° chisel edge angle, 8° primary clearance angle, and 130° point angle) was proposed based on a combination of the minimum exit delamination and least amount of burr height, predicted from Equations (6) and (7) to be 1.00528 and 82.2307 μm, respectively. The predicted optimized results for the responses Y_1 and Y_2 are tabulated in Table 7. For the suggested optimal drill bit shape, the discrepancies between the predicted and actual trial results were 0.11% and 9.72% respectively, hence validating that the proposed optimized cutter geometry was confirmed in the optimization model.

Table 7. Prediction of the optimized model of twist drill bit for edge defect analysis when drilling CFRP/Al7075-T6 stack-up material.

Responses (Y)	Y_1, [μm]	Y_2, [μm]
Model response	1.00528	82.2307
Experimental	1.00635	74.234
Error (%)	0.11	9.72

3.4. Hole Diameter Error

The hole accuracy attained here was in accordance with industry standards. This means the diameter tolerances fell within the range of the H8 zone, i.e., 18 μm [31,51]. The evaluation of the hole diameter error is shown in Figure 18 for both CFRP and Al7075-T6 plates. The absolute difference between the measured value and nominal value is the diameter error for CFRP and Al7075-T6. The variations in panel diameters between CFRP and Al7075-T6 are also noted and given the name "stack up diameter error".

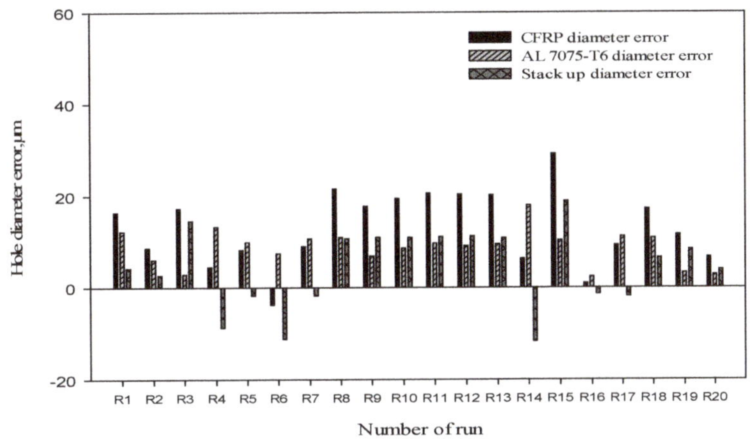

	R1	R2	R3	R4	R5	R6	R7	R8	R9	R10
$e_{a\,CFRP}$, [µm]	16.38	8.64	17.32	4.50	8.18	−3.76	8.96	21.56	17.78	19.48
$e_{a\,Al}$, [µm]	12.14	5.98	2.86	13.22	9.88	7.38	10.66	10.90	6.88	8.60
$e_{a\,Stack}$, [µm]	4.24	2.66	14.46	−8.72	−1.70	−11.14	−1.70	10.66	10.90	10.88
	R11	**R12**	**R13**	**R14**	**R15**	**R16**	**R17**	**R18**	**R19**	**R20**
$E_{a\,CFRP}$, [µm]	20.64	20.34	20.22	6.28	29.20	0.96	9.26	17.20	11.54	6.68
$e_{a\,Al}$, [µm]	9.62	9.12	9.46	17.96	10.28	2.36	11.10	10.72	3.23	2.78
$e_{a\,Stack}$, [µm]	11.02	11.22	10.76	−11.68	18.92	−1.40	−1.84	6.48	8.31	3.90

Figure 18. Hole diameter errors between stack-up materials for all runs.

According to the graph, the hole diameter of the CFRP material was found to be undersized only in R6 with a value of −3.76 µm. This is because a shrinking effect was induced by the drill geometry during the cutting operation [ref]. The cutting performance when drilling the CFRP panel would be decreased due to the narrower primary clearance angle. To limit the amount of shrinkage when drilling composite panels, a high clearance is required. The range for oversized holes in this research was 0.96 to 29.2 µm. Soo et al. [31] obtained a similar range of oversized holes between 6 to 34 µm while drilling CFRP/AA7010-T7451 with 6.38 mm flat point drills. Overall, R8, R10, R11, R12, R13, and R15 [35] did not meet the customer's specifications since one of the errors was greater than the OEM standard's permitted maximum. R16, the ideal cutter geometry, showed the least error for CFRP, Al7075-T6, and stack-up material. In all of the runs that were analyzed for stack-up error, values ranged from −1.40 to 18.92 µm.

When comparing individual hole diameters for CFRP and Al7075-T6, in some cases the hole diameters of Al7075-T6 were larger than those of CFRP and in some cases it was other way round. Both of these results were obtained in the past by various researchers, with different explanations. Soo et al. [31] mentioned that while drilling CFRP/AA7010-T7451 aluminum with a flat point drill, the Al layer hole was larger than the CFRP layer hole, as the former has a lower modulus of elasticity and higher thermal expansion compared to the latter [20]. Distinct elastic modulus values experience varying levels of elastic deformation during drilling, resulting in different hole dimensions. Additionally, when drilling Al7075-T6, the aluminum chips clogged the flute due to the specific geometry and raised the drilling temperature [52]. A smaller CFRP hole diameter is also due to the fact that the fibers flex back into the hole after a few days [53]. The cases where the hole diameter of CFRP was larger than that of Al7075-T6 are largely associated with continuous chip formation. Continuous chips likely twist along the drill body, leading to clogging. Thus, hot, sharp chips that are unable to be smoothly evacuated remain in the hole, enlarging the CFRP holes and deteriorating the surface quality of the CFRP.

3.5. Hole Circularity

Figure 19 displays the average hole circularity when drilling CFRP/Al7075-T6 stack material using various drill geometries. Overall, the average hole circularity of the CFRP laminates was better, ranging from 13.40 to 24.97 μm, compared to 13.23 to 26.07 μm for Al7075-T6. The results obtained here were better than the results obtained by ref. [31] in which drilling CFRP/AA7010-T7451 aluminum alloy gave values varying from 5 to 45 μm. In addition, the hole circularity error values obtained in the CFRP layer were generally larger than those in the Al7075 section [31], which was possibly due to tool runout causing radial deflection or initial chisel edge sliding (also known as 'walking') as the drill penetrated the top CFRP layer. No 'walking' was found in the current experiment, as no such trend in hole circularity errors was observed between CFRP and Al7075-T6.

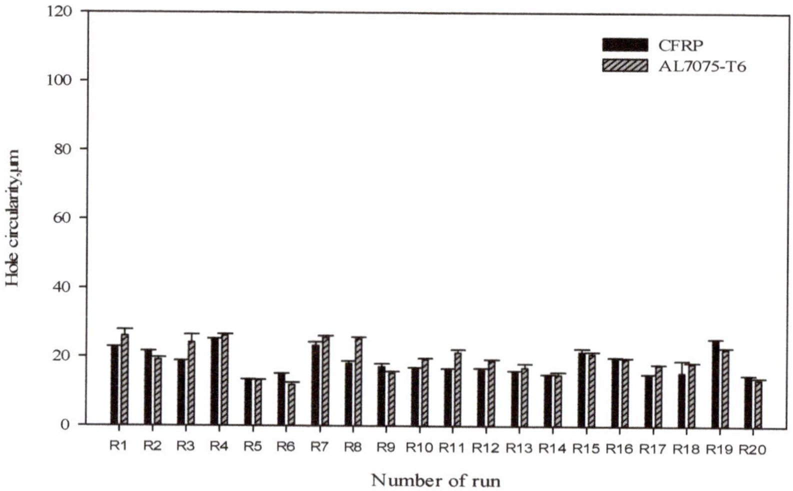

	R1	R2	R3	R4	R5	R6	R7	R8	R9	R10
Circularity$_{CFRP}$, [μm]	22.67	21.60	18.47	24.97	13.40	14.97	23.20	18.07	17.17	16.73
Circularity$_{Al}$, [μm]	25.97	19.10	24.13	26.07	13.23	11.83	25.53	25.10	15.27	19.23
	R11	**R12**	**R13**	**R14**	**R15**	**R16**	**R17**	**R18**	**R19**	**R20**
Circularity$_{CFRP}$, [μm]	16.30	16.27	15.70	14.60	21.40	19.70	14.60	15.50	25.03	14.57
Circularity$_{Al}$, [μm]	21.27	18.77	16.83	14.87	20.83	19.37	17.60	18.10	22.27	13.47

Figure 19. Hole circularity between stack-up materials for all runs in the extended study.

For all runs, the influence of the selected drilling parameter (2600 rev/min and 0.05 mm/rev) produced good results according to OEM standards. The smallest hole circularity error was found in R5 for CFRP (primary clearance angle = 8°, point angle = 140°, and chisel edge angle = 30°) and R6 for Al7075-T6 (primary clearance angle = 7°, point angle = 130°, and chisel edge angle = 37.5°). When the Al7075-T6 first encountered the drill bit when drilling the stack material, the stability was increased by reducing the tip angle to 130°. As a result, when the cutting tool needed to cut through the stack of material, there were less deflections and vibrations.

4. Conclusions

Detailed research, comprising experimentation, analysis, regression model construction, and optimization of the unique WC twist drill geometry, was successfully conducted to address hole edge defects and hole integrity of CFRP/Al7075-T6 stack-up material. The drilled hole in a CFRP panel can simply develop delamination while the drilled hole in an aluminum panel can be oversized without careful tool geometry and drilling parameter selection. As a result, the panel will be scrapped and the subsequent assembly procedure must be discontinued. Therefore, the ideal way to enhance the drilling process is to optimize tool geometry in order to boost drilling productivity and decrease the rejection of drilled parts.

- Although the point angle of the twist drill had to be raised from 130° to 140°, the delamination at the CFRP exit hole had a favorable effect on hole integrity.
- The average burr height was minimal, measuring between 40.2 and 271.2 μm. The lower burr height was found with a 45° chisel edge angle, 6° primary clearance angle, and 130° point angle. When the chisel edge angle was reduced to 30°, the burr height rose, resulting in a significant rolled-up phenomenon at the Al707-T6 panel's exit hole due to less available space for chip evacuation.
- The lowest hole diameter error values were obtained with values of 0.96 μm, 2.36 mm, and -1.4 μm for the stack-up diameter error, CFRP diameter error, and Al7075-T6 diameter error, respectively. At the same time, the hole circularity error was less than 30 μm in all runs, which was within OEM standards.
- Multiple response optimization was employed to optimize drill geometric parameters and the best drill geometry for a customized twist drill was proposed. To obtain minimal hole edge defects, it was discovered that the combination of 45° chisel edge angle, 8° primary clearance angle, and 130° point angle is the ideal drill geometry for a twist drill design, with a desirability index level of 0.773.

Author Contributions: Conceptualization, M.H.H. and J.J.M.; methodology, M.H.H.; validation, M.H.H. and J.X.; formal analysis, G.F.; investigation, G.F.; writing—original draft preparation, M.H.H. and J.J.M.; writing—review and editing, J.X., M.H.H. and G.F.; visualization, G.F.; supervision, M.H.H.; project administration, J.X. and M.H.H.; funding acquisition, J.X. and M.H.H. All authors have read and agreed to the published version of the manuscript.

Funding: This research was funded by a Short Term Grant (#304/PMEKANIK/6315557). The work was also partly funded by the National Natural Science Foundation of China (Grant Nos. 51705319 and 52175425).

Data Availability Statement: Not applicable.

Acknowledgments: The authors would like to acknowledge the support from Short Term Grant (#304/PMEKANIK/6315557) for financing this research work. They are also grateful for the financial participation of National Natural Science Foundation of China. Technical support from the School of Mechanical Engineering at USM is also greatly acknowledged.

Conflicts of Interest: The authors declare no conflict of interest.

References

1. Geng, D.; Liu, Y.; Shao, Z.; Lu, Z.; Cai, J.; Li, X.; Jiang, X.; Zhang, D. Delamination formation, evaluation and suppression during drilling of composite laminates: A review. *Compos. Struct.* **2019**, *216*, 168–186. [CrossRef]
2. Slamani, M.; Chatelain, J.-F. Assessment of the suitability of industrial robots for the machining of carbon-fiber reinforced polymers (CFRPs). *J. Manuf. Process.* **2019**, *37*, 177–195. [CrossRef]
3. Xu, J.; Zhou, L.; Chen, M.; Ren, F. Experimental study on mechanical drilling of carbon/epoxy composite-Ti6Al4V stacks. *Mater. Manuf. Process.* **2019**, *34*, 715–725. [CrossRef]
4. Li, C.; Xu, J.; Chen, M.; An, Q.; El Mansori, M.; Ren, F. Tool wear processes in low frequency vibration assisted drilling of CFRP/Ti6Al4V stacks with forced air-cooling. *Wear* **2019**, *426–427*, 1616–1623. [CrossRef]
5. Xu, J.; El Mansori, M.; Voisin, J.; Chen, M.; Ren, F. On the interpretation of drilling CFRP/Ti6Al4V stacks using the orthogonal cutting method: Chip removal mode and subsurface damage formation. *J. Manuf. Process.* **2019**, *44*, 435–447. [CrossRef]
6. Pecat, O.; Brinksmeier, E. Low damage drilling of CFRP/titanium compound materials for fastening. *Procedia CIRP* **2014**, *13*, 1–7. [CrossRef]
7. Pecat, O.; Brinksmeier, E. Tool wear analyses in low frequency vibration assisted drilling of CFRP/Ti6Al4V stack material. *Procedia CIRP* **2014**, *14*, 142–147. [CrossRef]
8. Zitoune, R.; Krishnaraj, V.; Collombet, F. Study of drilling of composite material and aluminium stack. *Compos. Struct.* **2010**, *92*, 1246–1255. [CrossRef]
9. Abdelhafeez, A.M.; Soo, S.L.; Aspinwall, D.K.; Dowson, A.; Arnold, D. Burr formation and hole quality when drilling titanium and aluminium alloys. *Procedia CIRP* **2015**, *37*, 230–235. [CrossRef]

10. Pilný, L.; De Chiffre, L.; Píška, M.; Villumsen, M.F. Hole quality and burr reduction in drilling aluminium sheets. *CIRP J. Manuf. Sci. Technol.* **2012**, *5*, 102–107. [CrossRef]
11. Liu, D.; Tang, Y.; Cong, W.L. A review of mechanical drilling for composite laminates. *Compos. Struct.* **2012**, *94*, 1265–1279. [CrossRef]
12. Vilches, F.J.T.; Hurtado, L.S.; Fernández, F.M.; Gamboa, C.B. Analysis of the chip geometry in dry machining of aeronautical aluminum alloys. *Appl. Sci.* **2017**, *7*, 132. [CrossRef]
13. Abhishek, K.; Datta, S.; Mahapatra, S.S. Optimization of thrust, torque, entry, and exist delamination factor during drilling of CFRP composites. *Int. J. Adv. Manuf. Technol.* **2015**, *76*, 401–416. [CrossRef]
14. Pawar, O.A.; Gaikhe, Y.S.; Tewari, A.; Sundaram, R.; Joshi, S.S. Analysis of hole quality in drilling GLARE fiber metal laminates. *Compos. Struct.* **2015**, *123*, 350–365. [CrossRef]
15. Luo, B.; Li, Y.; Zhang, K.; Cheng, H.; Liu, S. Effect of workpiece stiffness on thrust force and delamination in drilling thin composite laminates. *J. Compos. Mater.* **2016**, *50*, 617–625. [CrossRef]
16. Khashaba, U.A. Improvement of toughness and shear properties of multiwalled carbon nanotubes/epoxy composites. *Polym. Compos.* **2018**, *39*, 815–825. [CrossRef]
17. Geier, N.; Xu, J.; Pereszlai, C.; Poór, D.I.; Davim, J.P. Drilling of carbon fibre reinforced polymer (CFRP) composites: Difficulties, challenges and expectations. *Procedia Manuf.* **2021**, *54*, 284–289. [CrossRef]
18. Gaugel, S.; Sripathy, P.; Haeger, A.; Meinhard, D.; Bernthaler, T.; Lissek, F.; Kaufeld, M.; Knoblauch, V.; Schneider, G. A comparative study on tool wear and laminate damage in drilling of carbon-fiber reinforced polymers (CFRP). *Compos. Struct.* **2016**, *155*, 173–183. [CrossRef]
19. Khashaba, U.A.; El-Sonbaty, I.; Selmy, A.I.; Megahed, A.A. Machinability analysis in drilling woven GFR/epoxy composites: Part I—Effect of machining parameters. *Compos.-A Appl. Sci. Manuf.* **2010**, *41*, 391–400. [CrossRef]
20. Ashrafi, S.A.; Sharif, S.; Farid, A.A.; Yahya, M.Y. Performance evaluation of carbide tools in drilling CFRP-Al stacks. *J. Compos. Mater.* **2014**, *48*, 2071–2084. [CrossRef]
21. Wang, C.-Y.; Chen, Y.-H.; An, Q.-L.; Cai, X.-J.; Ming, W.-W.; Chen, M. Drilling temperature and hole quality in drilling of CFRP/aluminum stacks using diamond coated drill. *Int. J. Precis. Eng.* **2015**, *16*, 1689–1697. [CrossRef]
22. Heberger, L.; Kirsch, B.; Donhauser, T.; Nissle, S.; Gurka, M.; Schmeer, S.; Aurich, J.C. Influence of the quality of rivet holes in carbon-fiber-reinforced-polymer (CFRP) on the connection stability. *Procedia Manuf.* **2016**, *6*, 140–147. [CrossRef]
23. Lambiase, F.; Durante, M. Mechanical behavior of punched holes produced on thin glass fiber reinforced plastic laminates. *Compos. Struct.* **2017**, *173*, 25–34. [CrossRef]
24. Giasin, K. Machining Fibre Metal Laminates and Al2024-T3 Aluminium Alloy. Ph.D. Thesis, The University of Sheffield, Sheffield, UK, 2016.
25. Biermann, D.; Heilmann, M. Burr minimization strategies in machining operations. In *Burrs—Analysis, Control and Removal*; Aurich, J., Dornfeld, D., Eds.; Springer: Berlin/Heidelberg, Germany, 2010; pp. 13–20. [CrossRef]
26. Pilny, B.L. High speed drilling of aluminium plates. In *Faculty of Mechanical Engineering—Instituteof Manufacturing Technology*; BRNO Univresity of Technology: Brno-střed, Czech Republic, 2011; p. 151.
27. Ramulu, M.; Branson, T.; Kim, D. A study on the drilling of composite and titanium stacks. *Compos. Struct.* **2001**, *54*, 67–77. [CrossRef]
28. Ko, S.-L.; Lee, J.-K. Analysis of burr formation in drilling with a new-concept drill. *J. Mater. Process. Technol.* **2001**, *113*, 392–398. [CrossRef]
29. Sakurai, K.; Adachi, K.; Kawai, G.; Sawai, T.; Ogawa, K. High feed rate drilling of aluminum alloy. *Mater. Sci. Forum* **2000**, *331–337*, 625–630. [CrossRef]
30. Hamade, R.F.; Ismail, F. A case for aggressive drilling of aluminum. *J. Mater. Process. Technol.* **2005**, *166*, 86–97. [CrossRef]
31. Soo, S.L.; Abdelhafeez, A.M.; Li, M.; Hood, R.; Lim, C.M. The drilling of carbon fibre composite-aluminium stacks and its effect on hole quality and integrity. *Proc. Inst. Mech. Eng. B Manag. Eng. Manufact.* **2019**, *233*, 1323–1331. [CrossRef]
32. Mahdi, A.; Turki, Y.; Habak, M.; Salem, M.; Bouaziz, Z. Experimental study of thrust force and surface quality when drilling hybrid stacks. *Int. J. Adv. Manuf. Technol.* **2020**, *107*, 3981–3994. [CrossRef]
33. Benezech, L.; Landon, Y.; Rubio, W. Study of manufacturing defects and tool geometry optimisation for multi-material stack drilling. *Adv. Mat. Res.* **2011**, *423*, 1–11. [CrossRef]
34. Kuo, C.L.; Soo, S.L.; Aspinwall, D.K.; Thomas, W.; Bradley, S.; Pearson, D.; M'Saoubi, R.; Leahy, W. The effect of cutting speed and feed rate on hole surface integrity in single-shot drilling of metallic-composite stacks. *Procedia CIRP* **2014**, *13*, 405–410. [CrossRef]
35. Hassan, M.H.; Abdullah, J.; Franz, G. Multi-objective optimization in single-shot drilling of CFRP/Al stacks using customized twist drill. *Materials* **2022**, *15*, 1981. [CrossRef]
36. Rivero, A.; Aramendi, G.; Herranz, S.; Lopez de Lacalle, L.N. An experimental investigation of the effect of coatings and cutting parameters on the dry drilling performance of aluminium alloys. *Int. J. Adv. Manuf. Technol.* **2006**, *28*, 1–11. [CrossRef]
37. Kim, D.; Ramulu, M. Drilling process optimization for graphite/bismaleimide–titanium alloy stacks. *Compos. Struct.* **2004**, *63*, 101–114. [CrossRef]
38. Freddi, A.; Salmon, M. Introduction to the Taguchi method. In *Design Principles and Methodologies*; Springer Tracts in Mechanical Engineering; Springer: Cham, Switzerland, 2019; pp. 159–180. [CrossRef]

39. Behera, S.K.; Meena, H.; Chakraborty, S.; Meikap, B.C. Application of response surface methodology (RSM) for Optimization of leaching parameters for ash reduction from low-grade coal. *Int. J. Min. Sci. Technol.* **2018**, *28*, 621–629. [CrossRef]
40. Humbird, D.; Fei, Q. Chapter 20—Scale-Up Considerations for Biofuels. In *Biotechnologogy for Biofuel Production and Optimization*; Eckert, C.A., Trinh, C.T., Eds.; Elsevier Science: Amsterdam, The Netherlands, 2016; pp. 513–537. [CrossRef]
41. Şenaras, A.E. Parameter Optimization Using the Surface Response Technique in Automated Guided Vehicles. In *Sustainable Engineering Products and Manufacturing Technologies*; Elsevier: Amsterdam, The Netherlands, 2019; pp. 187–197. [CrossRef]
42. Sivaiah, P.; Chakradhar, D. Analysis and modeling of cryogenic turning operation using response surface methodology. *Silicon* **2018**, *10*, 2751–2768. [CrossRef]
43. Senthilkumar, M.; Prabukarthi, A.; Krishnaraj, V. Machining of CFRP/Ti6Al4V stacks under minimal quantity lubricating condition. *J. Mech. Sci. Technol.* **2018**, *32*, 3787–3796. [CrossRef]
44. Tsao, C.C.; Hocheng, H. Parametric study on thrust force of core drill. *J. Mater. Process. Technol.* **2007**, *192–193*, 37–40. [CrossRef]
45. Cai, B.-Y.; Ge, J.-P.; Ling, H.-Z.; Cheng, K.-K.; Ping, W.-X. Statistical optimization of dilute sulfuric acid pretreatment of corncob for xylose recovery and ethanol production. *Biomass Bioenergy* **2012**, *36*, 250–257. [CrossRef]
46. Guan, X.; Yao, H. Optimization of Viscozyme L-assisted extraction of oat bran protein using response surface methodology. *Food Chem.* **2008**, *106*, 345–351. [CrossRef]
47. Sui, S.; Song, G.; Sun, C.; Zhu, Z.; Guo, K.; Sun, J. Experimental investigation on the performance of novel double cone integrated tool in one-shot drilling of metal stacks. *Int. J. Adv. Manuf. Technol.* **2020**, *109*, 523–534. [CrossRef]
48. Gaitonde, V.N.; Karnik, S.R.; Achyutha, B.T.; Siddeswarappa, B. Taguchi optimization in drilling of AISI 316L stainless steel to minimize burr size using multi-performance objective based on membership function. *J. Mater. Process. Technol.* **2008**, *202*, 374–379. [CrossRef]
49. Hassan, M.H.; Abdullah, J.; Mahmud, A.S.; Supran, A. Burr height as quality indicator in single shot drilling of stacked CFRP/Aluminium composite. *Key Eng. Mater.* **2017**, *744*, 327–331. [CrossRef]
50. Siegel, A.F. (Ed.) Chapter 11—Correlation and Regression: Measuring and predicting relationships. In *Practical Business Statistics*, 7th ed.; Academic Press: Cambridge, MA, USA, 2016; pp. 299–354.
51. Giasin, K.; Hawxwell, J.; Sinke, J.; Dhakal, H.; Köklü, U.; Brousseau, E. The effect of cutting tool coating on the form and dimensional errors of machined holes in GLARE fibre metal laminates. *Int. J. Adv. Manuf. Technol.* **2020**, *107*, 2817–2832. [CrossRef]
52. Fernández-Pérez, J.; Cantero, J.L.; Díaz-Álvarez, J.; Miguélez, M.H. Hybrid composite-metal stack drilling with different minimum quantity lubrication levels. *Materials* **2019**, *12*, 448. [CrossRef]
53. Hassan, M.H.; Abdullah, J.; Mahmud, A.S.; Supran, A. Effect of drill geometry and drilling parameters on the formation of adhesion layer in drilling composite-metal stack- up material. *J. Mech. Eng.* **2018**, *5*, 90–98.

Journal of
Composites Science

MDPI

Article

Uniaxial Compressive Behavior of AA5083/SiC Co-Continuous Ceramic Composite Fabricated by Gas Pressure Infiltration for Armour Applications

Achuthamenon Sylajakumari Prasanth [1], Vijayan Krishnaraj [2,*], Jayakrishnan Nampoothiri [2], Ramalingam Sindhumathi [2], Mohamed Raeez Akthar Sadik [2], Juan Pablo Escobedo [3] and Krishna Shankar [3]

1 Department of Mechanical Engineering, PSG College of Technology, Coimbatore 641004, Tamilnadu, India; asp.mech@psgtech.ac.in
2 Department of Production Engineering, PSG College of Technology, Coimbatore 641004, Tamilnadu, India; jkn.prod@psgtech.ac.in (J.N.); sindhumathipsg@gmail.com (R.S.); mra.raeez@gmail.com (M.R.A.S.)
3 School of Engineering and Information Technology, The University of New South Wales, Canberra, ACT 2600, Australia; j.escobedo-diaz@unsw.edu.au (J.P.E.); k.shankar@adfa.edu.au (K.S.)
* Correspondence: vkr.prod@psgtech.ac.in

Abstract: A novel approach of a gas pressure infiltration technique is presented for the synthesis of Co-Continuous Ceramic Composite (C4). SiC foams of varying pore sizes were infiltrated with aluminium AA5083. Optical examination revealed that the SiC foams contained open cells with a network of triangular voids. The number of pores-per-inch (PPI) in the foams was found to depend on the strut thickness and pore diameter. The compressive strengths of two foam configurations, 10 and 20 PPI, were estimated to lie between 1–2 MPa. After infiltration, the compressive yield strength of the resulting C4 was observed to increase to 126 MPa and 120 MPa, respectively, for the 10 and 20 PPI C4. Additionally, the infiltration of ceramic foam with the AA5083 alloy resulted in an increase in strength of 58–100 times when compared with plain ceramic foam. The failure modes of the composites in compression were analyzed by crack propagation and determining the type of failure. The study revealed that shear failure and vertical splitting were the predominant mechanisms of compression failure, and that the fabricated C4 is advantageous in mechanical properties compared to the plain ceramic foam. This study, therefore, suggests the use of C4 composites in armour applications.

Keywords: co-continuous ceramic composites; C4; ceramic foam; gas infiltration; compressive strength; structural characterization

Citation: Prasanth, A.S.; Krishnaraj, V.; Nampoothiri, J.; Sindhumathi, R.; Akthar Sadik, M.R.; Escobedo, J.P.; Shankar, K. Uniaxial Compressive Behavior of AA5083/SiC Co-Continuous Ceramic Composite Fabricated by Gas Pressure Infiltration for Armour Applications. *J. Compos. Sci.* **2022**, *6*, 36. https://doi.org/10.3390/jcs6020036

Academic Editor: Jinyang Xu

Received: 31 December 2021
Accepted: 17 January 2022
Published: 20 January 2022

1. Introduction

Ballistic protection systems such as armour generally consist of several layers of materials. Each layer performs a specific role in attenuating the energy of projectiles [1,2]. Typically, a hard material such as ceramic is positioned on the front striking face and a matrix composite or high strength steel is placed as the backing face. The front face plate retards the striking force of the projectile by actions such as tumble, erosion, and fracture, whereas the backing material absorbs the residual kinetic energy of the projectile to bring the fragments to rest [3,4]. Ballistic protection plates composed of ceramic sticking face with polymer matrix composite (PMC) back plates have been extensively explored and proven for typical ballistic impact conditions. The abrasion resistance and hardness of the ceramic front face enables it to blunt the approaching projectile and absorb its energy to reduce the impact hazard. However, poor ductility limits its potential to take multiple hits. In addition, the processing complexities, cost, and weight of these systems restrict their wide use. Energy absorption studies in both quasi-static and dynamic conditions have revealed that steel–steel composite metal foams are suitable for armour. However, strict desiderata of lightweight materials for the strategic movement of defence personnel and vehicles restrain the use of such materials [5,6]. Hence, a review of prior studies indicates that weight reduction of ballistic protection plates is paramount in the research of alternative materials for armour.

A study by Chang et al. [6] reported the development of lighter bullet proof material composed of ceramic faced metal–ceramic interpenetrating composites (IPCs). These IPCs exhibited better impact resistance and were less susceptible to abrupt demolition due to the development of stress wave at the interface. The primary reason was attributed to acoustic impedance mismatch between the metal and ceramic. The interlocking type microstructure of the IPCs, also termed as co-continuous ceramic composites (C4), imparts improved fracture toughness, wear resistance, stiffness, and reduced distortions [7,8]. This distinctive combination of enhanced properties of C4 makes it suitable for applications which require high specific modulus, high strength, higher corrosion resistance, and improved abrasion properties, such as brake discs and armour [9,10]. Additionally, these features of C4 can originate new pathways in the design and fabrication of monocoque armour plates exhibiting appreciably less weight with better ballistic protection that may not be attained through a conventional approach. Placing emphasis on weight reduction, which is a requisite for the selection of high specific strength materials in aerospace and armour applications, previous studies have focused on the fabrication of aluminium-based composites by strengthening them with high-strength and rigid particulate ceramics such as B_4C, SiC, Si_3N_4, TiB_2 and TiC [10,11]. In C4 composites, the proportion of the ceramic phase can be higher than that of particle-reinforced composites due to their interpenetrated structure. Among the available set of ceramics, SiC is the preferred choice due its mechanical properties, ease of availability, and economic factors. Porous SiC preforms which are interconnected in three dimensions and impregnated with Al alloys, therefore, have the potential for applications in transportation and armour [12,13]. Nong et al. [14] reported the fabrication of 3D SiC/Al co-continuous composite to produce a ventilated shaft disc brake. The wear and friction behaviour of the prepared C4 in this study was comparable to that of cast iron and steel. In addition, better thermal conductivity and better wear resistance were obtained at half the density [15]. Bahrami et al. [16] fabricated bilayer Al/B_4C/rice husk ash composites by the pressureless infiltration method. The study revealed that the two factors namely, initial preform porosity and chemical composition of infiltrated alloys, exerted a substantial influence on the electrical resistivity and coefficient of thermal expansion, respectively. Pressureless infiltration was also utilized to fabricate Al/Si_3N_4 silica composites by Soltani et al. [17]. The results depicted that the processing temperature significantly influenced the modulus of elasticity of the composite. Recent studies by Prasanth et al. [18,19] report that gravity infiltration of SiC co-continuous foam with AA7075 and Al 6063 alloy is effective in enhancing the wear and toughness of the prepared C4 when compared to a monolithic infiltrant material. The compressive strength of such IPCs, produced by squeeze casting, was reported to be 660 MPa. This is higher than that of traditional composites reinforced with SiC particles [20]. Similarly, in another study, infiltration of Ni_3Al alloy into porous aluminium oxide by gas pressure infiltration was investigated. Composites with a low volume of Ni_3Al showed a fracture strength of 400 MPa. The highest volume fraction of Ni_3Al (30 vol.%) displayed a higher fracture strength of 675 MPa [21]. Among the diverse Al alloys, AA5083 is a potential alloy for the regime of C4 composites. Nevertheless, a systematic assessment of AA5083 as an infiltrant material to produce C4 is essential. Though the pressureless infiltration method [22,23] is economical and has the benefits of easy industrialization, pressure infiltration [24,25] is preferred for the manufacture of C4 due to its lower infiltration time, performance, and efficiency. A recent study by Zhang et al. [26] reported that mechanical-pressure infiltration method was effective for producing Al_2O_3/Al based materials for scaffoldings. The study also reported that the pressure infiltration technique enhanced the interfacial bonding between Al_2O_3 layers and the infiltrant Al alloy. Though C4 composites have been extensively analyzed for wear applications, studies with an emphasis on their compressive behavior for armour applications are scarce, to the best of the authors' knowledge.

Therefore, the focus of this investigation is on the synthesis, microstructural characterization, and a detailed analysis of the compressive behavior of AA5083/SiC C4 composites manufactured through the gas pressure infiltration technique. The properties of the constituent ceramic foam were also analyzed and discussed in comparison with the C4.

2. Materials and Methods

The materials utilized to manufacture the C4 composites are delineated in Sections 2.1 and 2.2 hereunder. The method of infiltration employed to manufacture the C4 is detailed in Section 2.3. Sections 2.4 and 2.5 describe the analyses of microstructural and mechanical properties respectively, conducted on the SiC foams and the C4 composites.

2.1. Ceramic Foam

In this study, SiC ceramic foams consisting of two different pore sizes, namely 10 and 20 pores per linear inches (ppi), were employed as the reinforcement network. The foams were manufactured by the replica method and commercially sourced from M/s Eltech Ceramics, Tamilnadu, India. Prior to fabrication of the C4, the chemical composition and purity of the SiC foams were characterized by X-ray diffraction (Empyrean Malvern Panalytical, Malvern, UK). Copper K-α X-rays of 1.5406 Å were utilized and the diffractions were collected at a scanning rate of 2°/min. The collected diffractogram was indexed and the peak intensity was used to determine the chemical composition of the SiC foams.

2.2. Infiltrant Alloy

The present study utilizes a commercially sourced AA5083 alloy from Disha Steels, Maharashtra, India in order to infiltrate the pores of the SiC foam to create the C4. The chemical composition of the as-received alloy, shown in Table 1, was analyzed using Optical emission spectroscopy (AMETEK-SPECTROMAXx, Kleve, Germany).

Table 1. Chemical composition of AA5083.

Elements \ Sample	Mg	Mn	Fe	Si	Cr	Cu	Zn	Ti	Al
Actual values	4.43	0.55	0.24	0.13	0.094	0.02	0.02	0.058	Balance
Nominal values	4–4.9	0.4–1	0.4	0.4	0.05–0.25	0.1	0.25 max	0.15 max	Balance

2.3. Manufacturing of C4

In this study, two configurations of C4 were manufactured by low pressure infiltration of AA5083 into the pores of the 10 and 20 ppi SiC foams respectively. A custom-designed Ar gas based set-up, as depicted in Figure 1, was utilized to perform the low pressure infiltration. The set-up comprises a pressure chamber of dimensions Ø100 mm × 450 mm capable of heating to a maximum temperature of 1200 °C and withstanding a maximum pressure of 6 bar. The pressure chamber was designed such that it served as a crucible for the dual purpose of melting and infiltration. The crucible containing SiC foam was preheated to 750 °C and a measured quantity of aluminum AA5083 ingot slices were added for melting. This specific sequence aided in reducing thermal shock exerted on the ceramic foam due to the difference between room temperature and the furnace atmosphere. It further prevented sudden clogging of molten Al during the infiltration process. Subsequently, the melt was superheated to 795 °C to accelerate the kinetics of infiltration. To facilitate pressure infiltration, the atmospheric air present in the chamber was evacuated by a rotary vacuum pump to a level of 10^{-2} bar. Subsequently, Ar inert gas was purged into the sealed chamber to raise the pressure to 4 bar. The composite melt was maintained in this environment for one hour to ensure complete infiltration. Thereafter, the infiltrated composite was allowed to solidify and cool to 400 °C in the furnace and eventually, by air cooling. The fabricated C4 was then machined out by a series of facing, turning and grinding operations.

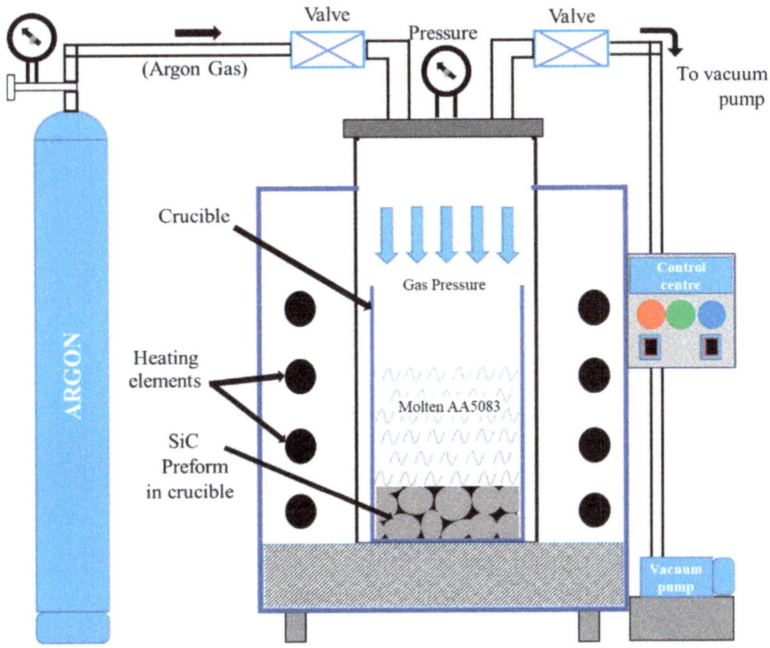

Figure 1. Gas pressure infiltration setup.

2.4. Analyses of Microstructural Properties

Subsequent to manufacturing the C4, the quality of infiltration in both configurations of SiC foam and C4 samples was determined by measuring the porosity levels using Archimedes' principle. The average porosity of three foam samples for each of the two configurations was determined using Equations (1)–(3).

$$\text{Porous Density, PD} = \frac{\text{Mass of porous foam}}{\text{Volume of porous foam}}\ \text{g/cc} \tag{1}$$

$$\text{Bulk Density, BD} = \frac{\text{Aggregrate mass of SiC particles in the foam}}{\text{Volume of porous foam}}\ \text{g/cc} \tag{2}$$

$$\text{Percentage Porosity, PP} = \left(1 - \frac{\text{PD}}{\text{BD}}\right) \times 100 \tag{3}$$

The porous density (PD) of the foam in Equation (1) represents the density encompassing the pores in the foam structure. The bulk density (BD) in Equation (2) was estimated using the actual quantity of SiC particles that constitute the foam. The percentage porosity (PP) in Equation (3) is a measure of the air pores available to be completely filled by AA5083 during the infiltration process. The volume fractions of the ceramic and Al phases in the C4 were estimated by Archimedes' principle using Equations (4) and (5).

$$\text{Volume fraction of Al in C4} = \frac{\text{Volume of Al in C4}}{\text{Volume of C4}} \tag{4}$$

$$\text{Volume fraction of SiC in C4} = \frac{\text{Volume of SiC in C4}}{\text{Volume of C4}} \tag{5}$$

Next, microstructural studies were performed on the two SiC foam configurations and the C4 composites thus manufactured. All metallographic samples analyzed in this study, were prepared by standard metallographic sample preparation procedures using a

Struers Tegramin-25 grinding and polishing machine. The optical micrographs of un-etched samples were captured using a ZEISS Axio Imager M2m optical microscope (OM).

2.5. Analyses of Mechanical Properties

The mechanical properties of SiC foams and the C4 were assessed through compression tests. A set of three samples each from SiC foams and C4 of dimensions Ø70 mm × 22 mm and Ø13 mm × 25 mm, respectively, were utilized as compression specimens. The compression tests were performed in accordance with ASTM E9 standard using a universal testing machine (Tinius Olsen, Norway) of capacity 50 kN with a cross head velocity of 0.5 mm/min. Subsequently, fractography analysis was performed in order to substantiate the observations from compression tests. The fractured face and the lateral sides of the specimen were examined using an optical microscope (Dino-lite capture) for indications of crack lines, shear failure, and vertical splitting failure of composite.

3. Results and Discussion

3.1. Metallographic Analysis of SiC Foams

The X-Ray diffraction pattern was extracted for the two configurations considered in this study, namely the 10 ppi (F10) and 20 ppi (F20) SiC foams. The identified constituents of the F10 and F20 foams and crystallographic structures along with their Joint Committee on Powder Diffraction Standards (JCPDS) reference patterns are listed in Table 2.

Table 2. SiC foam compounds identified by X-ray diffraction.

Foam Type	JCPDS	Compound	Crystallographic Structure	Distance between Atomic Planes (d)	Bragg Angle (2θ)	Miller Indices (hkl)
10 ppi (F10) and 20 ppi (F20)	00-029-1128	SiC	Hexagonal	2.62 2.50 2.35 2.17 1.54 1.40 1.30	34.08 35.65 38.15 41.40 59.99 65.70 71.70	101 102 103 104 110 109 202
	01-076-7775	Al_2O_3	Rhombohedral	3.40 2.55 2.38 2.09 1.60 1.40 1.37	25.51 35.06 37.68 43.24 57.35 66.34 68.03	012 104 110 113 116 214 300
	01-089-1304	$CaCO_3$ + Mg	Rhombohedral	3.02 1.90 1.86	29.50 47.70 48.66	104 018 116
	01-076-0940	SiO_2	Tetragonal	4.07	21.81	101

Next, during quantitative phase analysis, a refinement was performed with the High Score Plus 4.8. The results of the quantitative analysis of phase weight fractions and the percentage of each constituent for F10 is shown in Figure 2.

Figure 3 presents the indexed diffraction pattern for F20 along with the identified phases. All major peaks were assigned to the dominant phases, namely SiC and Al_2O_3. In addition, small peaks corresponding to $CaCO_3$ and SiO_2 were also identified. The compositional details of each compound are listed in Figure 3.

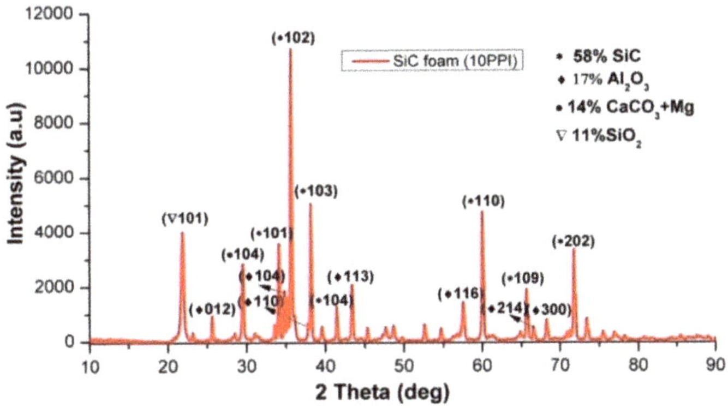

Figure 2. XRD pattern of F10.

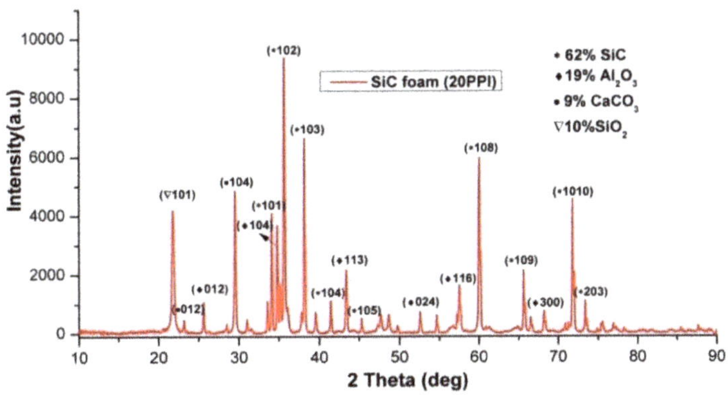

Figure 3. XRD pattern of F20.

The peak intensities of F10 and F20 as depicted in Figures 2 and 3 were used to calculate the weight fraction of each phase in a mixture and are tabulated in Table 3. The X-ray diffractograms confirmed that SiC was the major constituent of both the foams. In addition, compounds such as Al_2O_3, $CaCO_3$, Mg and SiO_2 were also observed.

Table 3. Weight fraction of phases.

Foam Configuration	SiC (%)	Al_2O_3 (%)	$CaCO_3$ + Mg (%)	$CaCO_3$ (%)	SiO_2 (%)
10 ppi	58	17	14	-	11
20 ppi	62	19	-	9	10

It can be inferred from Table 3 that the SiC content of F20 is higher than that of F10. It is well known that the ceramic phase, namely SiC, is brittle in nature. This can potentially lead to crack initiation and brittle fracture in the ceramic phase when subjected to compression [27].

3.2. Porosity and Structural Analysis of SiC Foams

The porosity of the foams was estimated using Equations (1)–(3) and is tabulated in Table 4. It can be inferred from Table 4 that the BD and PP of F20 is less than that of F10. This suggests a variation in the morphology of SiC struts between the two foam configurations. Therefore, a detailed study of the morphology of the two foam structures was conducted.

Table 4. Porosity of foams.

Foam Configuration	PD, g/cc	BD, g/cc	PP, %
10 ppi	0.477	2.476	80.73
20 ppi	0.469	2.062	77.25

Figure 4a–d represents the morphologies of the F10 and F20 SiC foams respectively The strut thickness and pore diameters of both foams are tabulated in Table 5.

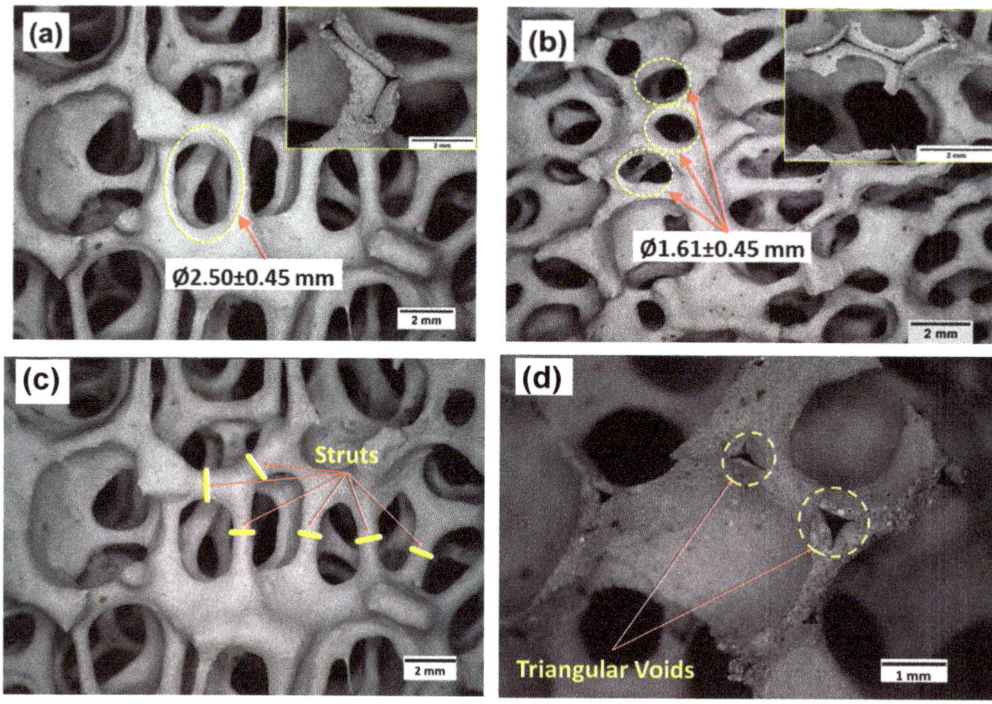

Figure 4. Optical Morphologies of the SiC foams: (**a**); 10 ppi; (**b**) 20 ppi; (**c**) struts; and (**d**) triangular voids in foam.

Table 5. Comparison of foam morphologies.

Foam Configuration	Strut Thickness (mm)	Pore Diameter (mm)
10 ppi (F10)	0.83 ± 0.25	2.5 ± 0.45
20 ppi (F20)	0.53 ± 0.14	1.61 ± 0.45

It can be inferred from Table 5 that the strut thickness and pore diameter of F20 is less than that of F10. Figure 4b reveals that F20 is highly interconnected when compared with F10. This signifies that a foam with lower strut thickness and pore diameter will possess a highly interconnected network of SiC. Additionally, it can be deciphered from Table 5 that the pore diameter decreases with an increase in the ppi of the foam.

In addition to the pore diameter and strut thickness, the strength of struts is a crucial factor influencing the mechanical strength of the foam. The SiC foams considered in this study consisted of an open-cell structure with a network of voids. These foams, as shown in the insets of Figure 4a,b, are termed as reticulated ceramics [28]. The extreme porosity, interconnected void volume, adjacent pores and their light weight make reticulated SiC ceramic foams ideal for the infiltration of molten metal [29]. As evident from Figure 4c,d, these porous structures possess hollow triangular voids that led to a reduction in their mass. Investigation of the strut structure of the F10 and F20 foams revealed that the estimated average side length of the triangular voids was ~466 μm and ~377μm, respectively.

3.3. Microstructural Analysis of C4

The optical micrographs of the as-cast C4 samples comprising the F10 and F20 foams infiltrated with AA5083 are shown in Figure 5. The ability to bear the compressive load exerted on the composite substantially depends on the morphology of the foam and infiltrant Al [20]. It can be observed that both infiltrated composites exhibit a spherical morphology. In particular, the C4-F10 exhibited a single lobe structure when compared with C4-F20 which reveals a double lobe structure. This distinction can be attributed to the small pore size of the F20 as detailed in Section 3.2. During formation of the composite, the molten Al impregnates and fills the pores of the foam, resulting in the characteristic lobe structures of the F10 and F20 foams. These distinctive morphologies, namely the lobe structures, have a salient effect on the compressive load-bearing capacity of the resulting C4 composite. Figure 5 also depicts through interpenetration of the Al alloy inside the voids of both SiC foams. The volume fractions estimated using Equations (4) and (5) are listed in Table 6. The table denotes that both configurations of C4 have approximately 80% by volume of AA5083 infiltrated into the SiC foams.

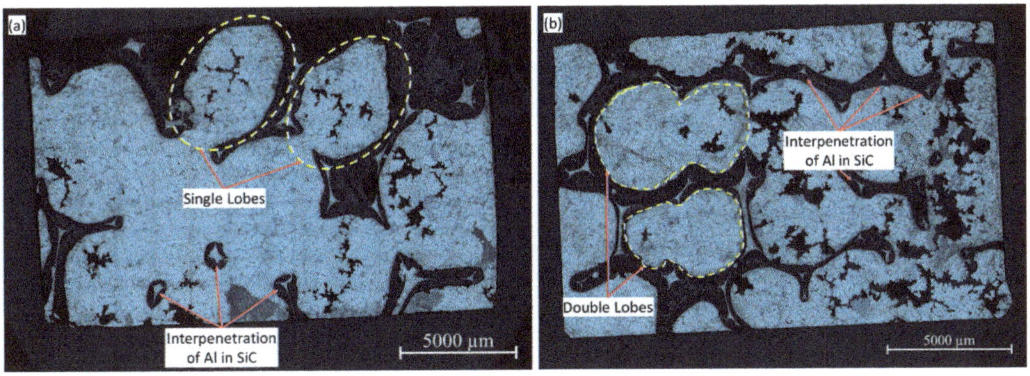

Figure 5. Microstructure of as-cast composite samples: (**a**) C4-F10; and (**b**) C4-F20.

Table 6. Volume Fraction of the C4.

Composite	**Volume Fraction (%)**	
	AA5083	**SiC**
10 ppi (C4-F10)	80.73	19.27
20 ppi (C4-F20)	77.25	22.75

3.4. Mechanical Property Analysis

As delineated in Section 2.5, the assessment of mechanical properties was performed by conducting compression tests on the two foam configurations, namely F10 and F20, and on the manufactured co-continuous composites, namely C4-F10 and C4-F20. The quasi-static stress-strain response of the SiC foams and three samples each of the C4 composites are depicted in Figure 6. The inferences of the salient outcomes from the compression tests are tabulated in Table 7.

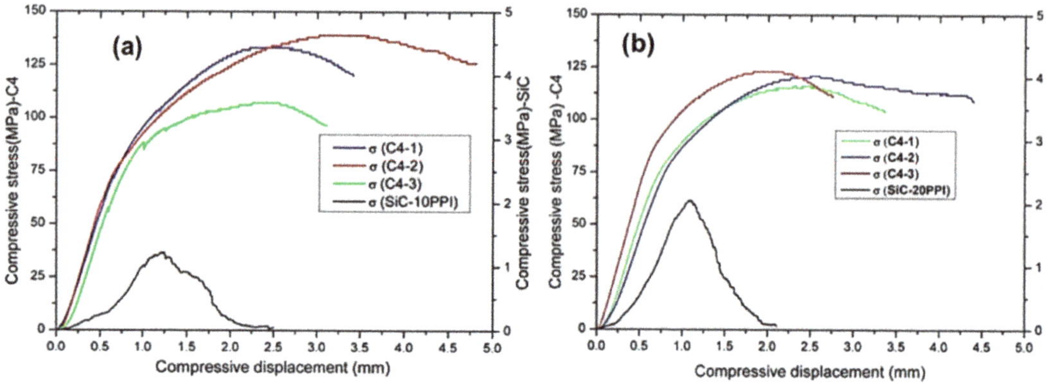

Figure 6. Compressive stress-strain behaviors of (**a**) 10 ppi SiC foam and its C4 (**b**) 20 ppi SiC foam and its C4.

Table 7. Inferences from Compression Tests.

	10 ppi		20 ppi	
	Foam (F10)	**C4 (C4-F10)**	**Foam (F20)**	**C4 (C4-F20)**
Elastic modulus	~0.96 MPa	~2.67 GPa	~2.3 MPa	~2.69 GPa
Yield strength (MPa)	~1	~74.3	~1.3	~71.6
Compressive strength (MPa)	~1.22	~126	~2.05	~120
Improvement in compressive strength	~100 times		~58 times	
Energy absorbed per unit volume (J/mm^3)	~1.07	~14.17	~1.68	~13.39
Improvement in energy absorption	~13 times		~8 times	

In the displacement curves in Figure 6, the strength of the C4 and SiC foam are marked on the primary and secondary *Y* axis respectively. From Table 7, the yield strength of F10 and F20 are observed to be ~1 MPa and ~1.3 MPa, respectively. In comparison, the displacement curves of the C4 show that the infiltration of AA5083 melt into the pores of SiC imparts an enhancement in yield strength with values of ~74.3 MPa and ~71.6 MPa for C4-F10 and C4-F20 samples, respectively. As can be deciphered from Table 7, the compressive strength of the SiC foam for the 10 and 20 PPI configuration was ~1.22 MPa and ~2.05 MPa. In contrast, the compressive strength of the C4-F10 composites was ~126 MPa and that of the C4-F20 configuration was ~120 MPa. It can be inferred that the compressive strength of 10 PPI SiC foam was enhanced by about 100 times by infiltrating with ~81 vol.% of AA5083 alloy. In comparison, the compressive strength of the 20 PPI SiC foam improved by close to 58 times when infiltrating with ~77 vol.% of AA5083 alloy.

Table 7 also lists the energy absorbed (EA) per unit volume by the foam and the C4 during compression tests. It was observed that the EA of SiC foam was estimated to be ~1.07 J/mm^3 and ~1.68 J/mm^3 for 10 PPI and 20 PPI foams respectively. The infiltration of SiC foams with AA5083 enhances the energy absorbed per volume of the C4 samples in both 10 and 20 PPI configurations to ~14.17 and ~13.39 J/mm^3, respectively. In addition to the strength and EA, the infiltration of AA5083 alloy remarkably enhanced the elastic modulus from ~0.96 and ~2.3 MPa for the foams to ~2.67 and ~2.69 GPa for the equivalent C4 composites as observed from Table 7. Therefore, it is evident that the infiltration of AA5083 into SiC ceramic foam simultaneously enhances the strength, elastic modulus and toughness, quantified by EA, of the resulting C4.

A salient inference from Table 6 is that the volume fraction of AA5083 in both 10 and 20 PPI composites are nearly the same. However, the compressive strength and EA of the composite is higher in case of the C4-F10 when compared with C4-F20, indicating that the compressive strength depends on the characteristics of the foam. The foam characteristics include PPI, pore diameter and strut thickness.

3.5. Compressive Behavior of C4

In order to comprehend the behaviour of the C4 when subjected to compression, a typical stress–strain curve obtained for the C4-F10 is shown in Figure 7. The curve can be segregated into four regions. The smooth line in region AB represents the elastic zone of the composite, during which no cracks were observed. Region BC represents the transition from elastic to failure zone. This was characterized by a minor bend in the curve with a reduced slope when compared to region AB. No cracks or strut failures occurred in the region AC. Subsequently, when the C4 traverses the transition region of BC, failure initiates after point C. It was observed during compression tests that the failure of the composite initiated within the foam was due to its brittle nature. The saw-tooth pattern in region CD was therefore attributed to the progressive cracking of the SiC struts. This was characterized by a cracking noise during testing. In this region CD, the AA5083 in the composite bears the load until point D. Finally, in region DE, the saw- tooth pattern was found to occur due to the splitting of the SiC-Al interface of the C4.

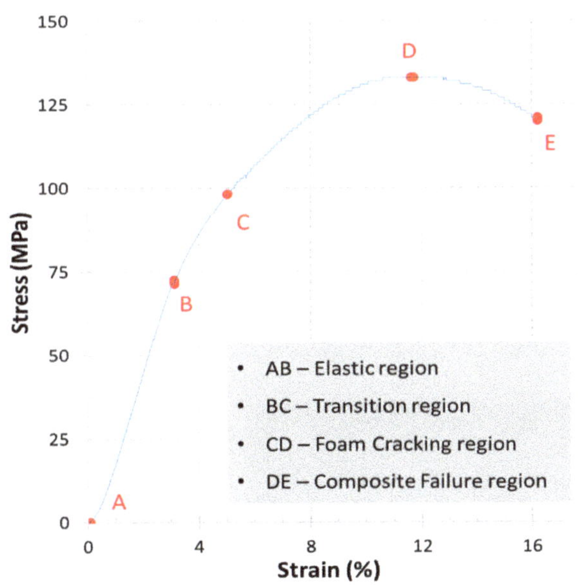

Figure 7. Compressive behavior of C4-F10.

3.6. Fractography Analysis

Analysis of the fractured surfaces, referred to as fractography, is essential to reveal the genesis and mechanism of failure [30]. Figure 8a shows the macroscopic image of the specimen of C4 before application of the uniaxial compressive load. It is evident from the figure that the AA5083 (bright phase) has infiltrated the pores of the SiC foam (dark phase) to form the C4. During compression tests, it was observed that, cracks first initiate at the SiC foam and then propagate to the ductile Al phase as evident from Figure 8b. Since both the C4-F10 and C4-F20 comprise approximately 80% of ductile Al and 20% of brittle SiC foam in a bulk form, it is expected to follow the 'shear failure' mode, characteristic of a ductile material. However, as evidenced from Figure 8b, the crack lines deviate from the theoretical line due to disruptions caused by SiC struts. Minor cracks of low intensity were observed to be formed and to propagate for short distances within the specimen. Although the 'shear failure' was expected, other types of failure such as 'Vertical splitting' were found to occur. Vertical splitting refers to failure along the direction in which the compressive load is applied [31]. This resulted in major cracks originating and propagating along the SiC struts, as deciphered from Figure 8c. The study of the fracture surfaces thus revealed that, the failure of C4 composites is contingent on the profile and orientation of the SiC struts. Hence, it can be proposed that, cracks initiate in the region of weak cell-walls of the foam structure. Subsequently, the cracks propagated to the surrounding cell-walls and plateau junctions. This correlates well with prior observations reported in literature [32–34].

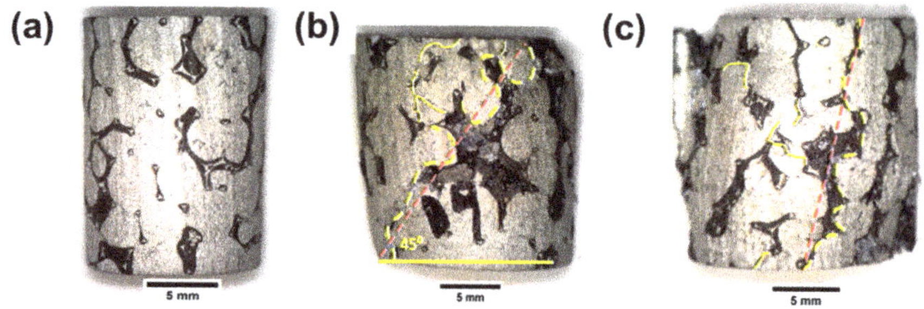

Figure 8. Macroscopic fractography of C4 composites: (**a**) composite without compression; (**b**) composite crack lines showing shear failure; and (**c**) composite crack lines showing vertical splitting failure.

3.7. Microstructure of Post-Test C4 Composite

Figure 9a,b exhibit the optical microstructure of the C4-F10 and C4-F20 respectively after compression tests. The microstructure reveals multiple cracks of varied sizes on the SiC struts in both configurations of the C4. In contrast, limited cracks are observed in the AA5083 matrix of C4 composite. This suggests that the quality of interfacial bonding between the Al matrix and foam structure is a paramount factor governing the compressive behavior of C4 composites. The phenomenon of shear failure and vertical splitting discussed in Section 3.6 is also evident from the micrographs.

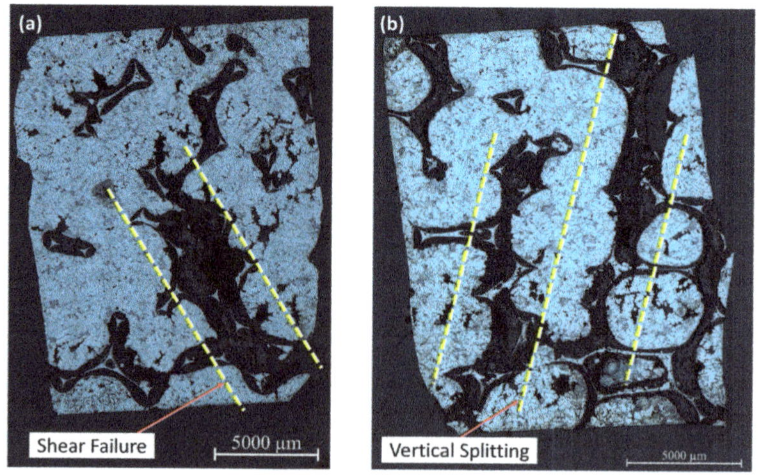

Figure 9. Optical micrograph of post-tested C4 composite (**a**) C4-F10 (**b**) C4-F20.

4. Conclusions

Two configurations of C4 composites composed of 81% and 77% by volume of AA5083 infiltrated into SiC ceramic foam of 10 and 20 PPI respectively, were synthesized using gas pressure infiltration technique. The foams and C4 were subjected to compression tests. The following conclusions were drawn from the study.

1. The XRD analysis for SiC foams (F10 and F20) revealed the presence of SiC, Al_2O_3, $CaCO_3$ and SiO_2. Additionally, F20 was found to possess higher amounts of brittle SiC when compared to F10.
2. The study of porosities indicated that the BD and PP of F20 was lower than that of F10. Extensive study of the morphology of the two foam structures revealed that the strut thickness and pore diameter of F20 is lower than that of F10. Additionally, F20 was observed to possess highly interconnected SiC strut structures when compared to F10.
3. The microstructure of the as-cast composite samples revealed single lobe and double lobe spherical structures for the C4-F10 and C4-F20, respectively. These characteristic lobe structures contribute to the compressive load bearing capacity of the composites.
4. The inference from compression tests was that, overall, the C4-F10 exhibited a better compressive strength of 126 MPa, a significant increase of nearly 100 times, when compared with the bare foam. This indicates that the characteristics of the chosen foam such as strut thickness, pore diameter, and the network of triangular voids is a crucial factor influencing the compressive strength of the C4.
5. The study also revealed that, infiltration of SiC foams with AA5083 enhanced the energy absorbed, strength and elastic modulus of the C4.
6. Fractography analysis revealed that cracks initiate in the frame of the C4 namely, the SiC foam structure. The AA5083 matrix delays the propagation of the cracks and thereby the premature failure of such composites.
7. Analysis of the compressive failure specimens indicated that the composites followed shear and vertical splitting failure modes. The orientation of the SiC struts was observed to be crucial in preventing crack propagation.

Author Contributions: Conceptualization, A.S.P., V.K., J.P.E. and K.S.; methodology, A.S.P., J.N., J.P.E. and K.S.; validation, J.N., J.P.E. and K.S.; formal analysis, R.S. and M.R.A.S.; investigation, J.N.; resources, A.S.P., V.K. and J.N.; data curation, R.S. and M.R.A.S.; writing—original draft preparation, R.S. and M.R.A.S.; writing—review and editing, A.S.P. and J.N.; visualization, V.K. and M.R.A.S.; supervision, V.K., J.P.E. and K.S.; project administration, V.K.; funding acquisition, A.S.P., V.K., J.P.E. and K.S. All authors have read and agreed to the published version of the manuscript.

Funding: This research was funded by The Scheme for Promotion of Academic and Research Collaboration (SPARC), MHRD, India. Project code: P207.

Institutional Review Board Statement: Not applicable.

Informed Consent Statement: Not applicable.

Conflicts of Interest: The authors declare no conflict of interest.

References

1. Garcia-Avila, M.; Portanova, M.; Rabiei, A. Ballistic performance of a composite metal foam-ceramic armor system. *Procedia Mater. Sci.* **2014**, *4*, 151–156. [CrossRef]
2. David, N.V.; Gao, X.-L.; Zheng, J.Q. Ballistic resistant body armor: Contemporary and prospective materials and related protection mechanisms. *Appl. Mech. Rev.* **2009**, *62*, 50802. [CrossRef]
3. Medvedovski, E. Ballistic performance of armour ceramics: Influence of design and structure. Part 1. *Ceram. Int.* **2010**, *36*, 2103–2115. [CrossRef]
4. Chabera, P.; Boczkowska, A.; Morka, A.; Kędzierski, P.; Niezgoda, T.; Oziębło, A.; Witek, A. Comparison of numerical and experimental study of armour system based on alumina and silicon carbide ceramics. *Bull. Polish Acad. Sci. Tech. Sci.* **2015**, *63*, 363–367. [CrossRef]

5. Jerz, J.; Simancik, F.; Bortel, M.; Kubo, S.; Kovacik, J.; Banhart, J.; Fleck, N.A. The design of lightweight armour sheets. *Banhart J. Fleck NA Mortensen A. Cell. Met. Manuf. Prop. Appl. MIT-Verlag* **2003**, *2003*, 43–46.
6. Chang, H.; Binner, J.; Higginson, R.; Myers, P.; Webb, P.; King, G. Preparation and characterisation of ceramic-faced metal-ceramic interpenetrating composites for impact applications. *J. Mater. Sci.* **2011**, *46*, 5237–5244. [CrossRef]
7. Travitzky, N.A.; Shlayen, A. Microstructure and mechanical properties of Al_2O_3/Cu–O composites fabricated by pressureless infiltration technique. *Mater. Sci. Eng. A* **1998**, *244*, 154–160. [CrossRef]
8. Chang, H.; Higginson, R.; Binner, J. Microstructure and property characterisation of 3-3 Al (Mg)/Al_2O_3 interpenetrating composites produced by a pressureless infiltration technique. *J. Mater. Sci.* **2010**, *45*, 662–668. [CrossRef]
9. Lu, Y.; Yang, J.; Lu, W.; Liu, R.; Qiao, G.; Bao, C. The mechanical properties of co-continuous Si3N4/Al composites manufactured by squeeze casting. *Mater. Sci. Eng. A* **2010**, *527*, 6289–6299. [CrossRef]
10. Jimoh, A.; Sigalas, I.; Hermann, M. In Situ Synthesis of Titanium Matrix Composite (Ti-TiB-TiC) through Sintering of TiH2-B4C. *Mater. Sci. Appl.* **2012**, *3*, 30–35. [CrossRef]
11. Meti, V.K.V.; Shirur, S.; Nampoothiri, J.; Ravi, K.R.; Siddhalingeshwar, I.G. Synthesis, Characterization and Mechanical Properties of AA7075 Based MMCs Reinforced with TiB2Particles Processed Through Ultrasound Assisted In-Situ Casting Technique. *Trans. Indian Inst. Met.* **2018**, *71*, 841–848. [CrossRef]
12. Daehn, G.S.; Breslin, M.C. Co-continuous composite materials for friction and braking applications. *JOM* **2006**, *58*, 87–91. [CrossRef]
13. Jiang, L.; Jiang, Y.L.; Yu, L.; Su, N.; Ding, Y.D. Experimental study and numerical analysis on dry friction and wear performance of co-continuous SiC/Fe-40Cr against SiC/2618 Al alloy composites. *Trans. Nonferrous Met. Soc. China* **2012**, *22*, 2913–2924. [CrossRef]
14. Nong, X.D.; Jiang, Y.L.; Fang, M.; Yu, L.; Liu, C.Y. Numerical analysis of novel SiC 3D/Al alloy co-continuous composites ventilated brake disc. *Int. J. Heat Mass Transf.* **2017**, *108*, 1374–1382. [CrossRef]
15. Maleque, M.A.; Dyuti, S.; Rahman, M.M. Material Selection Method in Design of Automotive Brake Disc. *Proc. World Congr. Eng.* **2010**, *III*, 5.
16. Bahrami, A.; Soltani, N.; Pech-Canul, M.I.; Soltani, S.; González, L.A.; Gutiérrez, C.A.; Tapp, J.; Möller, A.; Gurlo, A. Bilayer graded Al/B4C/rice husk ash composite: Wettability behavior, thermo-mechanical, and electrical properties. *J. Compos. Mater.* **2018**, *52*, 3745–3758. [CrossRef]
17. Soltani, N.; Soltani, S.; Bahrami, A.; Pech-Canul, M.I.; Gonzalez, L.A.; Möller, A.; Tapp, J.; Gurlo, A. Electrical and thermo-mechanical properties of CVI- Si3N4porous rice husk ash infiltrated by Al-Mg-Si alloys. *J. Alloys Compd.* **2017**, *696*, 856–868. [CrossRef]
18. Achuthamenon Sylajakumari, P.; Ramakrishnasamy, R.; Palaniappan, G. Taguchi Grey Relational Analysis for Multi-Response Optimization of Wear in Co-Continuous Composite. *Materials* **2018**, *11*, 1743. [CrossRef] [PubMed]
19. Achuthamenon Sylajakumari, P.; Ramakrishnasamy, R.; Palaniappan, G.; Murugan, R. Multi-response Optimization of End Milling Parameters for Al-Zn-Mg/SiC Co-continuous Composite Using Response Surface Methodology. *Medziagotyra* **2019**, *25*, 471–477. [CrossRef]
20. Zhao, L.-Z.; Zhao, M.-J.; Hong, Y.A.N.; Cao, X.-M.; Zhang, J.-S. Mechanical behavior of SiC foam-SiC particles/Al hybrid composites. *Trans. Nonferrous Met. Soc. China* **2009**, *19*, s547–s551. [CrossRef]
21. Skirl, S.; Krause, R.; Wiederhorn, S.M.; Rodel, J. Processing and mechanical properties of Al2O3/Ni3Al composites with interpenetrating network microstructure. *J. Am. Ceram. Soc.* **2001**, *84*, 2034–2040. [CrossRef]
22. Bahrami, A.; Pech-Canul, M.I.; Soltani, N.; Gutiérrez, C.A.; Kamm, P.H.; Gurlo, A. Tailoring microstructure and properties of bilayer-graded Al/B4C/MgAl2O4composites by single-stage pressureless infiltration. *J. Alloys Compd.* **2017**, *694*, 408–418. [CrossRef]
23. Bahrami, A.; Pech-Canul, M.I.; Gutierrez, C.A.; Soltani, N. Effect of rice-husk ash on properties of laminated and functionally graded Al/SiC composites by one-step pressureless infiltration. *J. Alloys Compd.* **2015**, *644*, 256–266. [CrossRef]
24. Xiao, Q.; Jiang, G. Effect of Powder Composition on the Preparation of SiC foam/Al Co-continuous Phase Composites by In-situ Reactive Pressureless Infiltration. *IOP Conf. Ser. Earth Environ. Sci.* **2021**, *714*, 032014. [CrossRef]
25. Vijayan, K.; Ramalingam, S.; Raeez Akthar Sadik, M.; Prasanth, A.S.; Nampoothiri, J.; Pablo Escobedo-Diaz, J.; Shankar, K. Fabrication of Co-Continuous ceramic composite (C4) through gas pressure infiltration technique. In *Proceedings of the Materials Today: Proceedings*; Elsevier Ltd.: Amsterdam, The Netherlands, 2021; Volume 46, pp. 1013–1016.
26. Zhang, Q.; Dong, S.; Ma, S.; Hou, X.; Yang, W.; Zhang, Y.; Wu, G. Microstructure and compressive behavior of lamellar Al2O3p/Al composite prepared by freeze-drying and mechanical-pressure infiltration method. *Sci. Eng. Compos. Mater.* **2020**, *27*, 1–9. [CrossRef]
27. Li, G.; Zhang, X.; Fan, Q.; Wang, L.; Zhang, H.; Wang, F.; Wang, Y. Simulation of damage and failure processes of interpenetrating SiC/Al composites subjected to dynamic compressive loading. *Acta Mater.* **2014**, *78*, 190–202. [CrossRef]
28. Steinacher, M.; Žužek, B.; Jenko, D.; Mrvar, P.; Zupanič, F. Manufacturing and properties of a magnesium interpenetrating phase composite. *Stroj. Vestn. J. Mech. Eng.* **2016**, *62*, 79–85. [CrossRef]
29. Islam, M.A.; Kader, M.A.; Hazell, P.J.; Escobedo, J.P.; Brown, A.D.; Saadatfar, M. Effects of impactor shape on the deformation and energy absorption of closed cell aluminium foams under low velocity impact. *Mater. Des.* **2020**, *191*, 108599. [CrossRef]

30. Kar, K.K. *Composite Materials: Processing, Applications, Characterizations*; Springer: Berlin/Heidelberg, Germany, 2017; ISBN 9783662495148.
31. Huang, R.; Li, P.; Liu, T. X-ray microtomography and finite element modelling of compressive failure mechanism in cenosphere epoxy syntactic foams. *Compos. Struct.* **2016**, *140*, 157–165. [CrossRef]
32. Kader, M.A.; Brown, A.D.; Hazell, P.J.; Robins, V.; Escobedo, J.P.; Saadatfar, M. Geometrical and topological evolution of a closed-cell aluminium foam subject to drop-weight impact: An X-ray tomography study. *Int. J. Impact Eng.* **2020**, *139*, 103510. [CrossRef]
33. Schukraft, J.; Lohr, C.; Weidenmann, K.A. 2D and 3D in-situ mechanical testing of an interpenetrating metal ceramic composite consisting of a slurry-based ceramic foam and AlSi10Mg. *Compos. Struct.* **2021**, *263*, 113742. [CrossRef]
34. Shi, Y.; Li, S.; Sitnikova, E.; Cepli, D.; Koch, D. Experimental evaluation and theoretical prediction of elastic properties and failure of C/C-SiC composite. *Int. J. Appl. Ceram. Technol.* **2022**, *19*, 7–21. [CrossRef]

Journal of
Composites Science

Article

Organomorphic Silicon Carbide Reinforcing Preform Formation Mechanism

Evgeny Bogachev

JSC Kompozit, 4 Pionerskaya, Korolev, 141070 Moscow, Russia; eug-bogatchev@mail.ru; Tel.: +7-495-513-23-06

Abstract: Development of the organomorphic ceramic-matrix composites (CMCs), where the reinforcing preform is built using polymer fibers subject essentially to hot pressing, was motivated by a desire to obtain much higher structural uniformity as well as to reduce the number of the process steps involved in the production of CMCs. This paper addresses the peculiarities of the organomorphic silicon carbide preform formation process. Using X-ray phase analysis, tomography, mass and IR spectroscopy, and thermomechanical and X-ray microanalysis, both the properties of the initial fibers of polycarbosilane (PCS)—the silicon carbide fiber precursor—and their transformation in the preform while heated to 1250 °C under constant pressing at 10–100 kPa were studied. Analysis of the data obtained showed the organomorphic SiC preform relative density at a level of 0.3–0.4 to be ensured by self-bonding of the silicon carbide preform, resulting from the fact that during the low-temperature part of pyrolysis, easily polymerizing substances are released leaving a high coke residue, thus cementing the preform. Another possible factor of SiC framework self-bonding is the destruction of the polymer fibers during pyrolysis of various PCS preforms differing in their methylsilane composition (for example, dimethylsilane), where deposition of silicon carbide on the contacting fibers starts as early as at 450–500 °C.

Keywords: polycarbosilane (PCS); fiber; pyrolysis; self-bonding; silicon carbide; organomorphic preform (OP)

Citation: Bogachev, E. Organomorphic Silicon Carbide Reinforcing Preform Formation Mechanism. *J. Compos. Sci.* **2023**, *7*, 81. https://doi.org/10.3390/jcs7020081

Academic Editor: Jinyang Xu

Received: 30 December 2022
Revised: 21 January 2023
Accepted: 8 February 2023
Published: 15 February 2023

1. Introduction

At the core of ceramic matrix composites (CMCs) is their reinforcing system. The existing technology for fabricating the reinforcing system involves multiple stages. These stages include a number of operations, namely, the production of elementary organic fibers from the relevant precursor, whereupon they are subject to step-by-step pyrolysis as part of a multifilament (made of thousands of filaments) tow: manufacturing of a fabric or a tape to be impregnated with a binder to obtain the prepreg; molding of the preform by curing; and carbonization that involves high-temperature annealing. There are certainly other ways for obtaining the reinforcing system using the available mechanical binding methods (multi-dimensional weaving, needle punching, and braiding); however, the reinforcing frames obtained using the aforesaid methods show a relatively low density and cannot be used as a basis for structural composites without the use of a binder. The numerosity of the operational processes involved results in the high fabrication cost of the preform as well as the CMC itself.

A special feature that defines the CMCs obtained using conventional methods is their rather non-uniform structure that, for the time being, cannot match the structural uniformity of graphite, metal, and ceramics. This prevents us from expanding the scope of CMC application as an alternative to these materials. It seems that any efforts to solve this problem by modifying a conventional preform, one way or another, are bound to fail. The use of multifilament tows, which are immanent features of the state-of-the-art technology, will result intrinsically in a non-uniform distribution of the reinforcement in relation to the composite volume, since the filament diameter range between 7 μm and 15 μm, and the size of the voids between the multifilament tows will make up 300–400 μm to 700–1000 μm and between the filaments—0.3–0.7 μm. The notable micro non-uniformity of the reinforcing preform will impose a natural limitation on its minimum thickness. Obtaining a composite material where the elementary fibers alternate uniformly with the matrix, thus bringing the dimensional limit of the material essentially about the diameter of the fiber itself, would bring the lower limit of the thickness of the parts down to 150–200 μm. Such a possibility, along with a notable reduction in the number of the process steps, may expand significantly the scope of application of CMCs through inclusion of many promising applications where CMC can be used instead of graphite, traditional ceramics, or metal.

The search for a solution of this problem resulted in the development of the so-called organomorphic CMCs (C/C, C/SiC, and SiC/SiC) reinforced with a framework based on organic polymer fibers used for the most common carbon and ceramic fibers, namely, polyacrylonitrile (PAN), polycarbosilane (PCS), and polysilazane (PSZ) [1].

Using PAN fibers as an example, hot pressing (up to 1000 °C) of PAN preforms was found [2] to provide carbon reinforcing frames, showing a high relative density of up to 0.4 and an open porosity of 50–60% represented by pores of an equivalent diameter ranging from several micrometers to several tens of micrometers. The method used to build the reinforcing carbon preform was found to provide self-bonding (as inherited after pyrolysis) of the filaments. The new quality of the carbon preforms made it possible both to provide an effective substitute for the molybdenum electrodes used for ion thruster accelerating electrodes [3] and to find more effective methods to CVI them with the silicon carbide matrix [4].

Self-bonding during so-called hot pressing (up to 170 °C) is a well-known phenomenon typical of natural fibers such as cotton and wood [5–8]. This phenomenon, resulting from a physical contact between the fibers as well as the presence of a bonding-conducive polymer on their surface, makes it possible to fabricate high-strength boards made of natural materials without any binder.

However, the organomorphic CMC preforms made of polymer fibers are subject to annealing at up to 1800 °C, and they are not limited by the process temperatures characteristic of the natural fibers. Therefore, it is important to determine the presence of self-bonding in the PCS preforms pyrolized under pressing, including at the low-temperature (up to 400 °C) pyrolysis stage. Formerly, self-bonding was observed in pyrolized PAN-based preforms [1,2].

The PAN- and PCS-fibers differ significantly in their chemical composition and structure. At the same time, it seems that the self-bonding discovered previously for PAN fibers during pyrolysis under pressure must be basically of general nature, and the governing laws of obtaining the carbon and SiC preforms, respectively, must be more or less the same.

2. Materials and Methods

The PCS fibers of Kompozit JSC make, thermally stabilized at 200 °C (molecular mass distribution $M_n \geq 1600$, polydispersity coefficient $D \leq 1.8$, 90% wt. residue after etching treatment) in air, 19–24 μm in diameter, were studied with the Skyscan-2011 nanotomograph (London, United Kingdom) using the following scanning parameters: U = 50 kV, I = 200 mkA, rotation angle—0.3° as averaged over 5 frames, and spatial resolution—0.3 μm/pixel.

The PCS fiber behavior, while under load, was studied with the thermomechanical analysis method (TMA) using the TMA Q400 EM (New Castle, Delaware, USA). The test sample in the form of layers of fibers, 7 mm in diameter and 3.4 mm in thickness, was placed in a special pot made of high-density organomorphic C/C, 7 mm in diameter, 11 mm in height, and 1.5 mm in wall thickness. Next, the fiber sample was pressed with a quartz indenter, 2.54 mm in diameter (Figure 1), to be heated up to 400 °C in air at a rate of 1 deg/min, under a static load of 0.588 N and a dynamic load of (±0.294 N), keeping a record of the time change in the elasticity modulus and sample thickness.

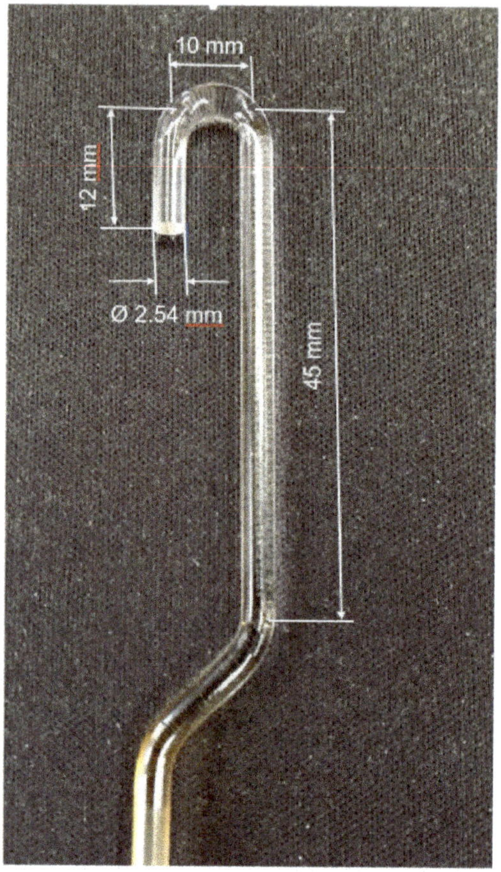

Figure 1. EXPANSION probe.

The structure and functional groups of the PCS fiber macromolecules, both before and after TMA, were analyzed using X-ray tomography with the XT H 320 LC X-ray tomography system (METRIS, Tring, UK) and using IR spectroscopy with the Netzsch STA 449 F3 Jupiter (Selb, Germany).

The PCS fiber thermal decomposition kinetics was studied in high-purity nitrogen flow with the TGA-DSC method using the TA Instruments SDT 650 synchronous differential thermal analyzer (New Castle, DE, USA) at the rate of a temperature rise of 10 degrees per minute.

The PCS fiber gas emission during heating up to 800 °C in vacuum was studied using the laboratory mass spectrometer equipped with the CIS300 quadrupole mass analyzer (Sunnyvale, CA, USA). The PCS fibers, 1–5 mg in weight, were placed in a quartz ampoule connected to the quadrupole mass spectrometer vacuum system. The ampoule was pumped out at the room temperature down to ~10^{-3} Pa and connected to the mass spectrometer chamber. While keeping continuous record of the mass spectra, the ampoule was subject to heating from the room temperature up to 800 °C at a constant rate of 5 deg/min. The mass spectra were recorded every 90 s, over the mass number range of 1–220. Simultaneously, record of the pressure in the mass spectrometer chamber (sensor MID-2) and in front of the mercury diffusion pump behind the nitrogen trap (sensor MID-1) was kept.

To obtain the organomorphic preforms from the silicon carbide fiber, the fibers in the form of a tape composed of 150-filament threads were placed in a graphite container with the internal dimensions of 120 × 60 × 28 mm in the directions of 0° × 0° and 0° × 90°, whereupon a commensurable massive graphite punch was installed onto the fibrous PCS workpiece, placing an additional load on the top [9]. The total load on the fibrous polymer workpiece (cover + load) was 700 N. The ready assembly was installed in a vacuum resistance furnace to be heated up to 1250 °C in non-oxidizing gas (nitrogen) at a rate of 300–400 deg/h. After cooling down, the organomorphic SiC preform was removed from the container to determine the density, whereupon test samples of about 20 × 10 mm in size were cut out of it with a sharp knife for microstructural analysis.

The density and porosity of SiC preforms were measured using the Archimedes method.

To study the structural changes in the material, the scanning electron microscope FEI Quanta 600 FEG (Eindhoven, The Netherlands) with field emission and integrated microanalysis system EDAX TRIDENT XM 4 was used. For X-ray phase analysis, the Empyrean diffractometer (PANalytical B.V., Willmington, Delawer, USA) equipped with the specialized HighScore Plus software for phase analysis with a built-in database of reference structures PAN-ICSD (Inorganic Crystal Structure Database) was used.

3. Results

3.1. Initial PCS Fiber Properties

Analysis of the initial PCS fibers reveals a completely amorphous and disordered nature of the structure, as evidenced by the fact that the small-angle scattering shows no reflection (Figure 2).

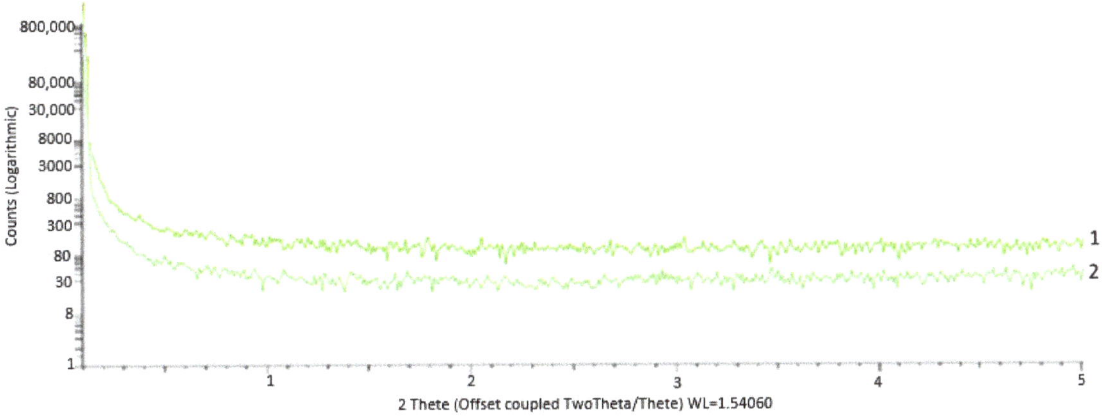

Figure 2. Meridional (1) and equatorial (2) scans of PCS fiber bundles in the small-angle area.

Microtomography of the polymer fibers shows no systematic difference in the near-surface and intra-volume areas (Figure 3).

It is clear, however, that there are more or less dense microvolumes in the fibers, which speaks for certain structural defects of the organic filaments.

3.2. Analysis of the PCS-Fiber Properties during the Pyrolysis

Thermomechanical analysis of the PCS fiber bundle 3-mm thick reveals complex deformation of the test sample under simultaneous static/dynamic loading with rising temperatures (Figure 4).

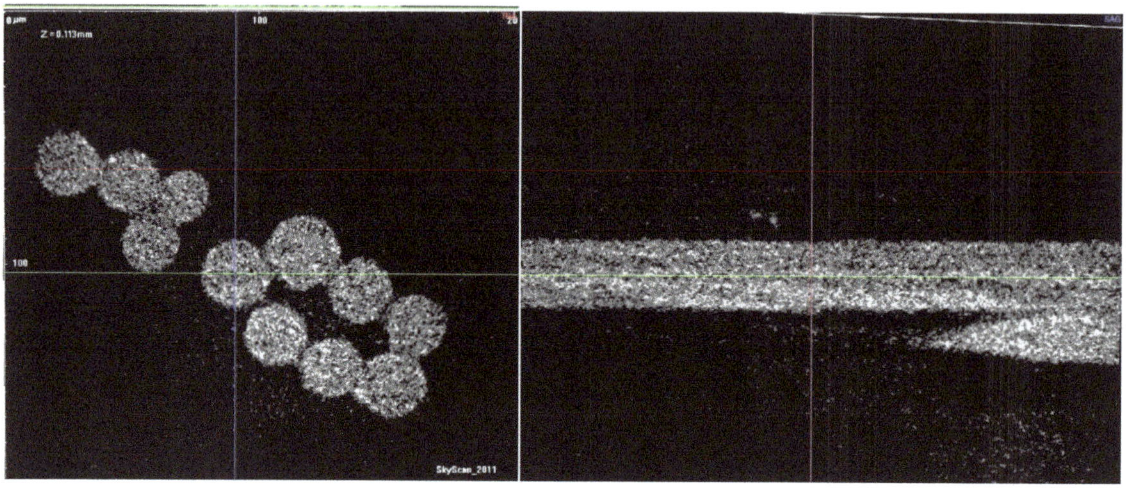

Figure 3. Interperpendicular sections of PCS fibers.

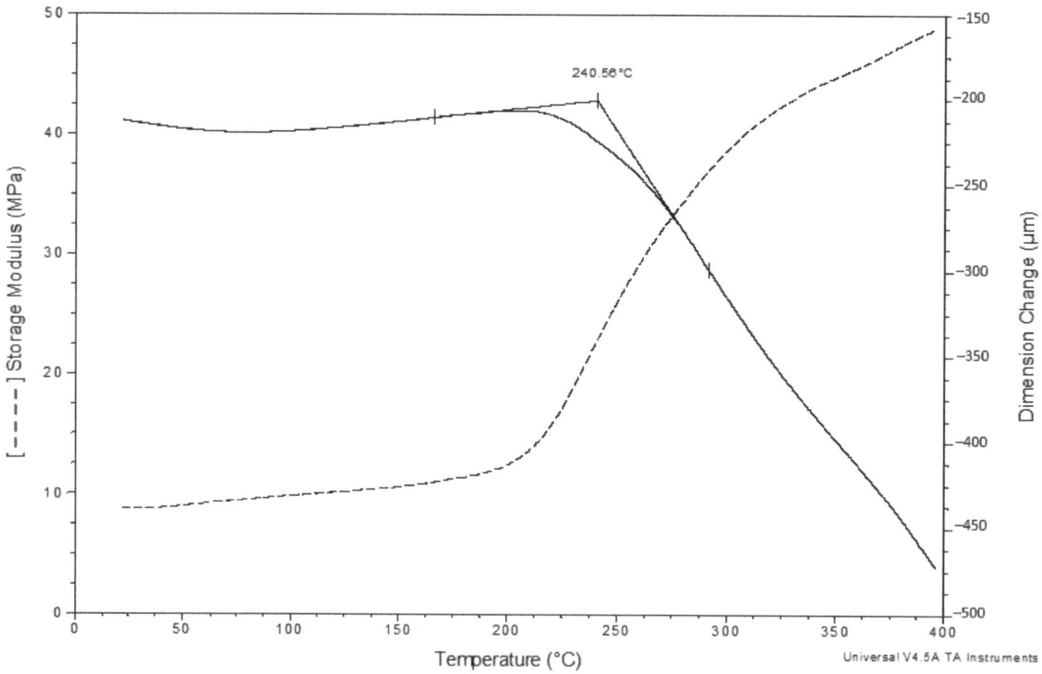

Figure 4. Elasticity modulus vs. temperature curves obtained for the PCS fiber bundle.

What draws attention is the timely coincidence of the sharp decrease in the sample thickness and the rapid increase in the elasticity modulus—both processes start at 215–220 °C. The decrease in thickness can be explained both by the partial tangential displacement of the fibers towards the pot walls with rising temperature due to friction reducing between the fibers and the increase in the degree of their contact simultaneously with the deformation of the fibers themselves at a temperature preceding the pyrolysis onset temperature. On the DTG curve, the onset of gas emission from PCS fibers practically corresponds to the start of the events during TMA (Figure 5).

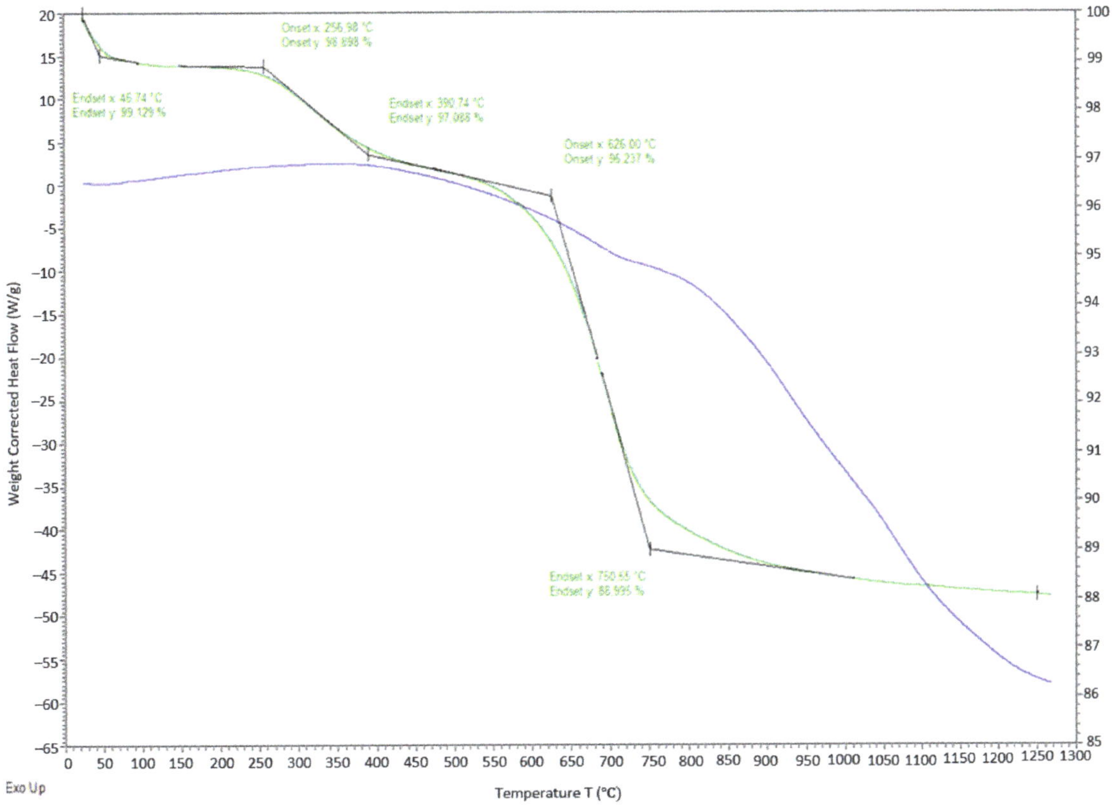

Figure 5. Thermogravimetric analysis of PCS fibers.

Mass spectroscopy of the PCS fibers shows that, according to the MID-2 pressure sensor readings, the low-temperature peak of gas emission starts at 215 °C, reaches its maximum at 270 °C, and ends at 470 °C (Figure 6).

The MID-1 pressure sensor fails to record this broad peak, meaning that in the specified temperature range, release of volatile substances with the mass-to-charge ratios of 18 (H_2O), 28 (CO), and 44 (CO_2) that condense at the liquid nitrogen temperature takes place. These low-molecular weight compounds ensure the loss of about 2% wt. of PCS fibers in the temperature range under study, i.e., 215–470 °C (Figure 5). Removal of oxygen-containing compounds from the composition of the fibers results inevitably in the partial breakage of the oxygen cross-link in PCS macromolecules, ensuring an increase in their mobility and ability for autohesive interaction.

The second (higher) peak of gas emission starts at 500 °C, reaches its maximum at 660 °C, and ends at 740 °C. In this case, the MID-1 pressure sensor does show it, meaning that it is associated mainly with release of hydrogen, since it is not frozen at the liquid nitrogen temperature. The mass spectrometry results also correlate with the DTA data of differential thermal analysis (Figure 5). However, according to [10], a significant mass loss (more than 7% wt.) is provided by the evolution of both hydrogen and methane.

Thus, in the range of 215–400 °C, transformations occur both in the composition and in the structure of the PCS fibers, leading both to a decrease in the thickness of the compressed sample and an increase in its elasticity modulus.

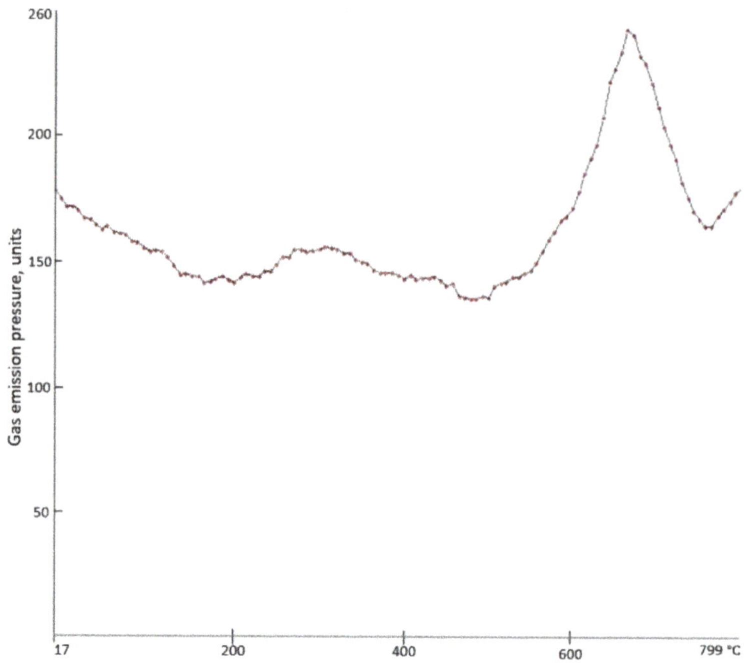

Figure 6. Changes in the total gas release from the PCS fibers when heated in the mass spectrometer.

3.3. Analysis of the PCS-Based Organomorphic Preform (OP) Microstructure and Properties after Different Stages of the Pyrolysis

After TMA, the test sample represents a fairly dense tablet (Figure 7).

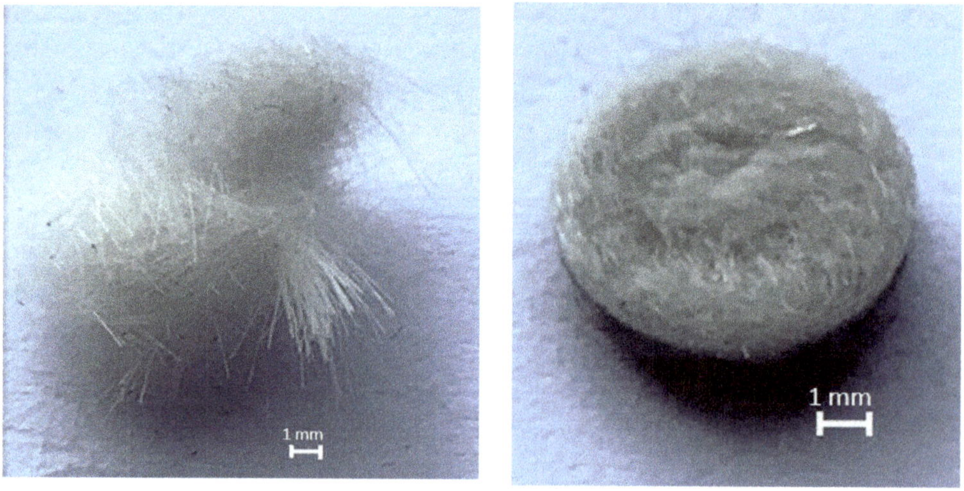

(**a**) (**b**)

Figure 7. PCS test sample before (**a**) and after (**b**) thermomechanical analysis at up to 400 °C.

Analysis of the PCS sample microstructure after TMA shows clearly visible insular buildups on the fiber surface (Figure 8), while in the initial state the fiber surface was free of them.

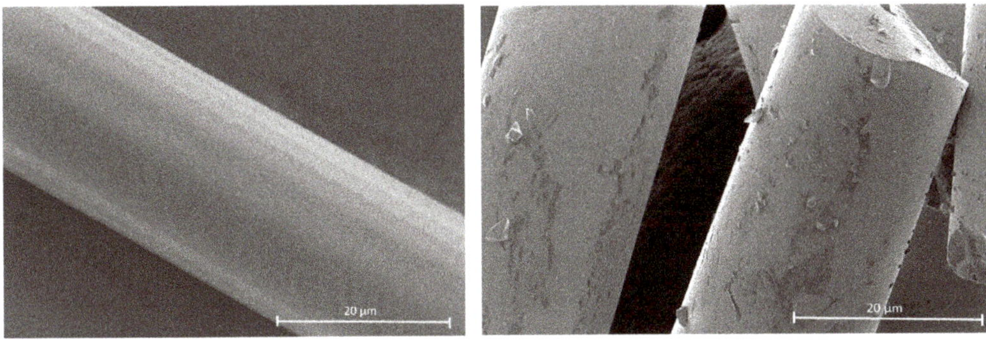

(a) (b)

Figure 8. Morphology of PCS fibers before (**a**) and after (**b**) thermomechanical analysis at up to 400 °C.

These buildups are not present on all fibers and are characterized by irregular concentrations. It is probably for that reason that, after TMA, the PCS sample shows less irreversible compression in the indenter pressure area than the PAN sample, the entire surface of which is covered with resin-forming compounds that effectively bind the filaments [2] (Figures 9 and 10).

Figure 9. Morphology of PAN fibers after pressing at 180 °C [2].

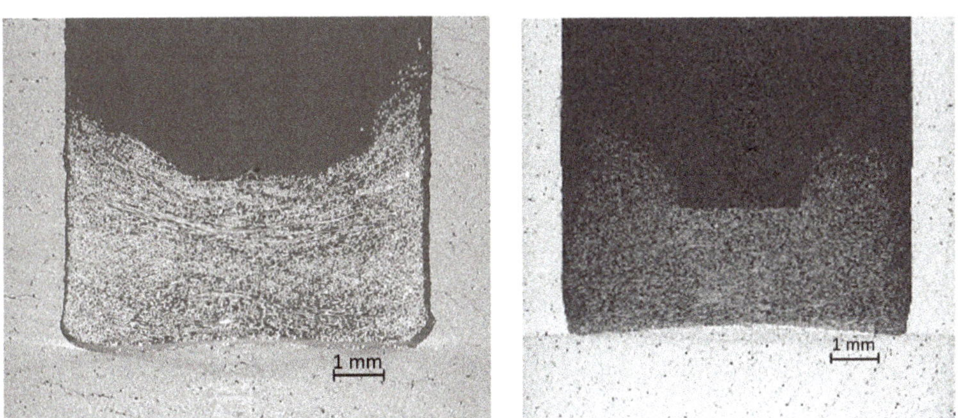

(a) (b)

Figure 10. Tomographic images after TMA: (**a**) PCS fiber and (**b**) PAN fiber [2].

Analysis of the PCS fiber sample composition using IR spectroscopy showed the sample underwent significant changes during TMA; what remains in the IR absorption spectrum of the fiber after TMA are signals at 780, 1220 cm^{-1} ($Si–CH_3$ bonds), 1440 cm^{-1} (C–H bonds), and 980 cm^{-1} ($Si-CH_2–Si$ bonds) [10,11] (Figure 11).

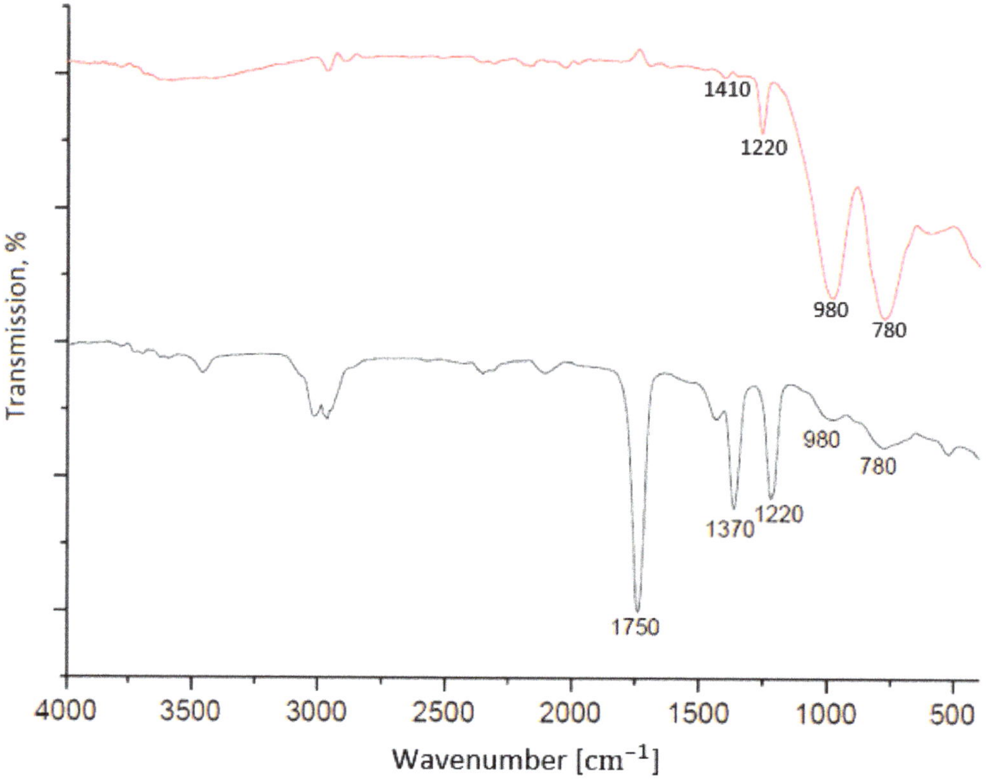

Figure 11. Results of IR-spectroscopy of the PCS fibers before (1) and after (2) TMA.

After TMA, signals at 1370, 1750, and 2960–3020 cm^{-1}, part of which are related to oxygen-containing bonds, disappear. These changes also correlate with both the results of mass spectroscopy and the compressed fiber transformation. The persistence of the irreversible deformation of the PCS sample after TMA, albeit to a lesser extent than in the case of the PAN sample (Figure 10), explains the increase in the elastic modulus of the compressed region during TMA; formations protrude onto the surface of the filaments and interact with each other autohesively to fix the gradual transformation of individual fibers into a single whole.

Therefore, the insular nature of the self-bonding areas of the fibers does not prevent the formation of dense and highly processible reinforcing silicon carbide frames of various reinforcement structures that proceed from organic to inorganic state, accompanied by their shrinkage as a whole (Figure 12).

(a)

(b)

(c)

Figure 12. OP-SiC: (**a**) unidirectional PCS tape in a container before pyrolysis; (**b**) SiC frame, 90 × 45 × 4 mm in size, reinforcement scheme 0°/0° [1]; and (**c**) SiC frame, 90 × 45 × 3.5 mm in size, reinforcement scheme 0°/90°.

Characteristics of the organomorphic silicon carbide preforms are given in Table 1.

Table 1. Characteristics of the organomorphic silicon carbide preforms.

Parameter	Value
Density, g/cm^3	0.9–1.0
Open/closed porosity, %	51–53/0
Fiber volume content in the preform, not less than, %	42
Fiber diameter in the preform, µm	15–17
Shrinkage in length, %	25–30
Shrinkage by fiber diameter, %	15–20
Mass loss during pyrolysis, %	19–20

The diffraction pattern obtained for the SiC frame after PCS preform pyrolysis shows two maxima: one of angular width 2θ of about 10° and the other of −20° (Figure 13).

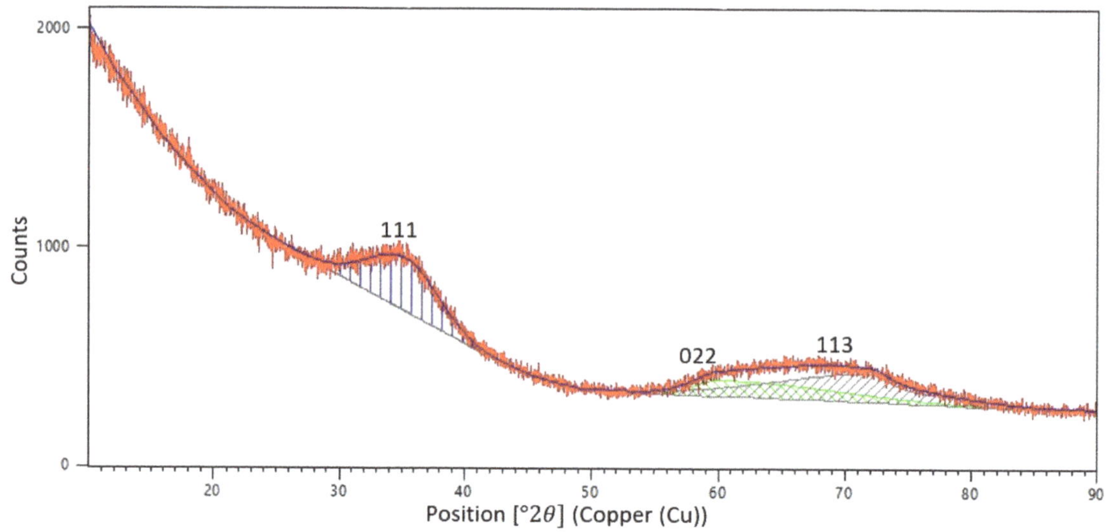

Figure 13. Diffraction pattern of the organomorphic SiC-preform (wide peak analysis was performed using the specialized HighScore Plus software for phase analysis).

Absence of narrow (angular width 2θ < 2°) diffraction peaks from the crystalline phase is indicative of a disordered—close to amorphous—state of the frame fibers. This is quite understandable, since the structure of the initial PCS fibers is completely amorphous (Figures 2 and 3) and the maximum temperature for obtaining the organomorphic SiC framework (1250 °C) does not reach even half of the incongruent decomposition temperature of silicon carbide [12]. However, the first—less wide—maximum corresponds to the strongest line (111) of β-SiC (reference code 98-002-8389), since this modification is formed during annealing of silicon carbide in a nitrogen atmosphere [13]. The second wide peak is apparently a superposition of other two strong lines, namely, (022) and (113).

Analysis of the silicon carbide preform microstructure confirms presence of some sort of bridges binding the fibers to each other (Figure 14).

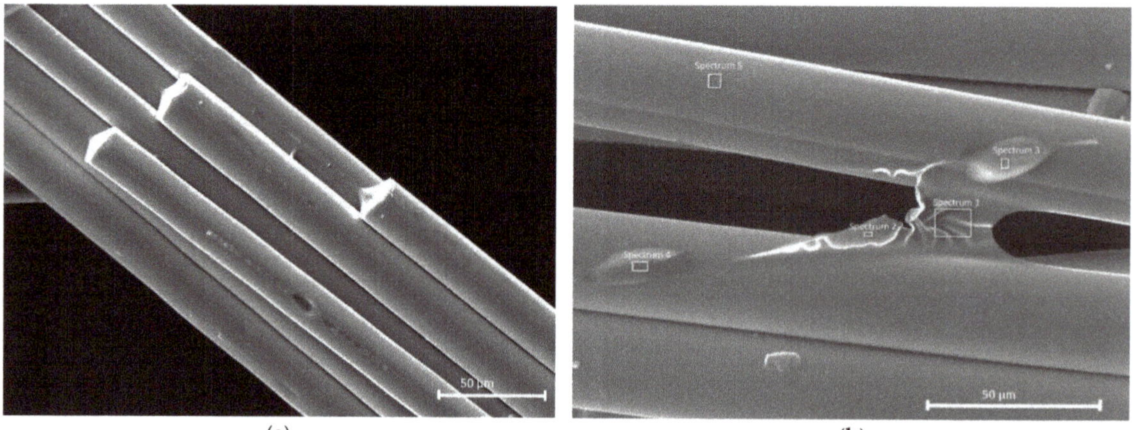

(a) (b)

Figure 14. SiC frame microstructure showing filament self-bonding areas (**a**,**b**). The boxes (spectrum 1 to 5) correspond to the areas for which the energy dispersive spectroscopy results were obtained.

According to energy dispersive spectroscopy results (Table 2), the chemical composition of the binding bridge (see Figure 14b; spectrum 1 and spectrum 2) can be represented approximately as Si:C:O = 1:13:1, which indicates enrichment with carbon.

Table 2. Energy dispersive spectroscopy results of the SiC-filaments within the preform after pyrolysis at 1250 °C.

Element	Content, % at.				
	Spectrum 1	Spectrum 2	Spectrum 3	Spectrum 4	Spectrum 5
Si	6.45	6.41	23.42	33.25	26.98
C	84.81	86.71	55.22	48.47	54.51
O	8.74	6.88	21.36	18.28	18.51

The carbon content decreases sharply in the buildups (spectrum 3, spectrum 4) and their composition practically corresponds to the composition of the surface of the fiber itself (spectrum 5). Since the morphology of the SiC fiber buildup is similar to the buildup's nature of the PCS fibers shown in Figure 8, one can suggest similarity of their origin. The excessive carbon in spectrum 1 and spectrum 2 suggests that the carbon polymer can prevail over the organosilicon one in the buildups and accretions. This reveals possible composition difference among the filaments coming out on the surface to form buildups that are composed of parts of PCS macromolecules in the preform under pressing during the low-temperature (up to 400 °C) part of pyrolysis.

4. Discussion

The results obtained place a new light on the processes occurring during PCS fiber pyrolysis. Annealing of loose fibers results in production of also loose silicon carbide fibers. Pyrolysis of the PCS fiber wound on a spool results in poor unwindability of the resulting SiC fibers, as during carbidization, the PCS fibers when shrinking, compress the underlying layers. Finally, PCS fiber pyrolysis under load produces a dense, highly processible silicon carbide framework. Thus, one should recognize that the main factor affecting self-bonding of the PCS fibers during organic-to-inorganic state transition is their compression while being heated. Gluing of the fibers that not only prevents their unwinding during pyrolysis in the wound state but also contributes to organomorphic frame cementation is the result of the formation of buildups or accretions on the fiber surface from the matter released from the PCS fibers during the low-temperature part of pyrolysis under compressive force.

According to [14], for the half-width of the projection of elastic deformation area r_o of two equal parallel cylinders of the same material (e.g., PCS fibers in a unidirectional preform), we have:

$$r_o = 1.52(qR/2E)^{1/2} \quad (1)$$

where q—specific load per unit fiber length in a unidirectional organomorphic frame; R—PCS fiber radius; and E—PCS fiber elasticity modulus.

During contact, for the maximum stress that develops along the axis of the contact area, we have:

$$\sigma_{max} = 0.418(2qE/R) \quad (2)$$

The maximum tangential stress $\tau(0, z)$ falls on the following depth:

$$z_o/r_o = 0.8 \quad (3)$$

and makes up about one third of σ_{max}.

Estimates made using Formulas (1)–(3) show that with the r_o value of about 4.5 µm, the maximum stress σ_{max} values make up approximately 0.03 MPa and the maximum tangential stress of about 0.01 MPa develops at a depth of 3.6 µm. In addition, it should be borne in mind that due to an uneven degree of the fiber contact as observed in the actual preforms, the stress values in some contact bundles may be significantly higher.

Thus, both at the contact area and at the depth comparable to the half-width of the contact area, well-marked normal and tangential stresses develop that may affect the state of PCS macromolecules in the polymer fibers during heating and its accompanying shrinkage.

As evidenced by the configuration of the bridge between the filaments (see Figure 14), deformation of the buildups may appear even in the form of a viscous flow. Moreover, the polymer coming out on the fiber surface may be capable of curing, which may cause certain changes in the nature and a sharp increase in the elasticity modulus at above 200 °C (see Figure 4). The other reason for modulus increase is the increase in the degree of fiber contact during TMA [15].

It is obvious that these buildups come out on the filament surface from the bulk and correlate with the fiber composition. Since thermal stabilization of the PCS fibers in air is of a diffusive nature, the parts of PCS macromolecules possessed of greater mobility may diffuse through the defects and discontinuities in the more cross-linked surface layer under stress-strain conditions and upon reaching the onset temperature of partial decomposition of oxygen crosslinks. Undoubtedly, they are capable of active autohesive interaction with the surface of adjacent filaments.

Variations in the buildup composition over a wide range of carbon, oxygen, and silicon content, as well as their different concentration on the PCS filament surface, are likely to be associated with different defects in the local areas of the fiber surface and with difference in the degree of stitching of their respective microvolumes. This makes different parts of the PCS macromolecules come out on the surface under stress-strain conditions. Another possible reason for formation of SiC frames is the release of various methylsilanes from the polymer fibers during pyrolysis of the PCS preforms [16], among which dimethylsilane $(CH_3)_2SiH_2$ was detected by mass spectroscopy. Decomposition of this compound accompanied by deposition of silicon carbide starts as early as 450–500 °C.

5. Conclusions

The carbon organomorphic preform formation mechanism studied earlier using PAN fibers as an example [2] showed that self-bonding of the organic fibers within the preform is due to the presence of a solid polymer film on the fiber surface and a continuous physical contact between the filaments during pyrolysis when combined with hot pressing.

The investigation into the organomorphic silicon carbide preform formation mechanism reveals similarity of its underlying laws with those inherent in the formation of carbon frames. However, the insular nature of the buildups composed of the matter released from the PCS fibers during the low-temperature (up to 400 °C) part of pyrolysis reduces the volume-wise continuity of self-bonding in the organomorphic SiC preform. All the while, the relative densities of the silicon carbide preforms are still higher than 40%.

Apparently, the fundamental similarity of the self-bonding nature of the PAN- and PCS-preforms allows us to formulate the following basic prerequisites for origination of the contacts between the fibers within the preforms:

1. Stabilized structure of the polymer fibers to exclude their melting during pyrolysis.
2. Chemical purity of the surface of the contacting polymer filaments (absence of any obstacles for autohesion, i.e., diffusive merging of the polymer fibers).
3. Sufficient continuous mechanical pressing of the fibers against each other during pyrolysis (a constant load).
4. Duration of the contact between the filaments, when in the polymer state, sufficient for the autohesion interaction to form.

However, the presumable deposition of a nanoscale SiC layer from dimethylsilane $(CH_3)_2SiH_2$ on the fiber surface may complicate, at a later stage, the task of obtaining high-strength SiC–SiC CMCs based on the silicon carbide organomorphic reinforcing framework.

Funding: This research received no external funding.

Institutional Review Board Statement: Not applicable.

Informed Consent Statement: Not applicable.

Data Availability Statement: All the data supporting the reported results can be found in this manuscript. No additional data are available in the publicly archived datasets.

Acknowledgments: The author extends his great appreciation to the colleagues who assisted in carrying out the research: I.O. Leipunsky, Head of the Laboratory of Nano- and Microstructural Materials Science at the Institute of Energy Problems of Chemical Physics of the Russian Academy of Sciences, D.V. Onuchin, Deputy Vice-Rector for Science at the Mendeleev University of Chemical Technology and D.M. Kiselkov, Senior Scientist at the Institute of Technical Chemistry UB of the Russian Academy of Sciences. All individuals included in this section have consented to the acknowledgement.

Conflicts of Interest: The author declares no conflict of interest.

References

1. Bogachev, E.A. Design, Structure and Properties of Organomorphic Composites as New Materials. *Ceram. Int.* **2019**, *45*, 9537–9543. [CrossRef]
2. Bogachev, E.A. Organomorphic carbon preform formation mechanism. *J. Compos. Sci.* **2022**, *6*, 50. [CrossRef]
3. Akhmetzhanov, R.V.; Balashov, V.V.; Bogachev, Y.A.; Yelakov, A.B.; Kashirin, D.A.; Svotina, V.V.; Spivak, O.O.; Cherkasova, M.V. An Ion Thruster Accelerating Electrode Made of Carbon-carbon Composite Material. *Therm. Eng.* **2018**, *65*, 986–993. [CrossRef]
4. Kulik, A.V.; Bogatchev, E.A.; Kulik, V.I. Experimental Study and Numerical Modeling of Pulse CVI for the Production of Organomorphic C/SiC Composites. *Int. J. Appl. Ceram. Technol.* **2021**, *19*, 91–100. [CrossRef]
5. Fahmy, T.; Mobarak, F. Advanced Binderless Board-like Green Nanocomposites from Undebarked Cotton Stalks and Mechanism of Self-bonding. *Cellulose* **2013**, *20*, 1453–1457. [CrossRef]
6. Wang, Q.; Xiao, S.; Shi, S.Q.; Cai, L. Mechanical Property Enhancement of Self-bonded Natural Fiber Material via Controlling Cell Wall Plasticity and Structure. *Mater. Des.* **2019**, *172*, 107763. [CrossRef]
7. Zhao, Y.; Yue, J.; Tao, L.; Liu, Y.; Shi, S.Q.; Cai, L.; Xiao, S. Effect of Lignin on the Self-bonding of a Natural Fiber Material in a Hydrothermal Environment: Lignin Structure and Characterization. *Int. J. Biol. Macromol.* **2020**, *158*, 1135–1140. [CrossRef] [PubMed]
8. Rowell, R.M.; McSweeny, J.D. Heat Treatments of Wood Fibers for Self-Bonding and Stabilized Fiberboards. *Mol. Cryst. Liq. Cryst.* **2008**, *483*, 307–325. [CrossRef]
9. Bogachev, E.A. Method for Fabricating a Porous Preform for Composite Materials. Patent of Russia No. 2685130, 16 April 2019.
10. Tsyrlin, A.M. *Continuous Inorganic fibers for Composites*; Metallurgy Publisher: Moscow, Russia, 1992; 237p.
11. Kuptsov, A.H.; Zhizhin, G.N. *Handbook of Fourier Transform Raman and Infrared Spectra of Polymers*; Technosphera Advertising & Publisher Center: Moscow, Russia, 2013; 696p.
12. Haynes, W.M. *CRC Handbook of Chemistry and Physics*, 92nd ed.; CRC Press: Boca Ration, FL, USA, 2011; p. 488.
13. Kawamura, T. Silicon Carbide Crystals Grown in Nitrogen Atmosphere. *Mineral. J.* **1965**, *4*, 333–355. [CrossRef]
14. Shtremel, M.A. *Destruction. Book 1*; Publisher House of the Moscow Institute of Steel and Alloys: Moscow, Russia, 2014; 670p.
15. Balshin, M.Y. *Scientific Foundations of Powder Metallurgy and Fiber Metallurgy*; Metallurgy Publisher: Moscow, Russia, 1972; 336p.
16. Qiu, H.P.; Chen, M.W.; Jiao, J.; Li, X.Q.; Wang, Y.; Gao, S.Y. Studies on the Pyrolysis Kinetic Pehaviours of Polycarbosilane. In Proceedings of the 8th International Conference on High Temperature Ceramic Matrix Composites (HT-CMC 8), Xi'an, China, 22–26 September 2013; p. 382.

Journal of
Composites Science

MDPI

Article

Optical Detection of Void Formation Mechanisms during Impregnation of Composites by UV-Reactive Resin Systems

Benedikt Neitzel [1,*] and Florian Puch [1,2]

[1] Plastics Technology Group, Department of Mechanical Engineering, Technische Universität Ilmenau, 98693 Ilmenau, Germany
[2] Thüringisches Institut für Textil- und Kunststoff-Forschung e.V., 07407 Rudolstadt, Germany
* Correspondence: benedikt.neitzel@tu-ilmenau.de; Tel.: +49-3677-69-3345

Abstract: During the impregnation of reinforcement fabrics in liquid composite molding processes, the flow within fiber bundles and the channels between the fiber bundles usually advances at different velocities. This so-called "dual-scale flow" results in void formation inside the composite material and has a negative effect on its mechanical properties. Semi-empirical models can be applied to calculate the extent of the dual-scale flow. In this study, a methodology is presented that stops the impregnation of reinforcement fabrics at different filling levels by using a photo-reactive resin system. By means of optical evaluation, the theoretical calculation models of the dual-scale flow are validated metrologically. The results show increasingly distinct dual-scale flow effects with increasing pressure gradients. The methodology enables the measurability of microscopic differences in flow front progression to validate renowned theoretical models and compare simulations to measurements of applied injection processes.

Keywords: void formation; dual-scale flow; permeability; textile preforms; liquid composite molding; fiber reinforced plastics

Citation: Neitzel, B.; Puch, F. Optical Detection of Void Formation Mechanisms during Impregnation of Composites by UV-Reactive Resin Systems. *J. Compos. Sci.* **2022**, *6*, 351. https://doi.org/10.3390/jcs6110351

Academic Editor: Jinyang Xu

Received: 2 October 2022
Accepted: 11 November 2022
Published: 15 November 2022

1. Introduction

Liquid composite molding (LCM) is an established industrial production technology for manufacturing thermoset fiber composites. Dry textile preforms are draped in a mold, which is subsequently closed, and the semi-finished fiber product is impregnated with a resin system by means of overpressure. During impregnation of the textile preforms, voids are formed, which result in a reduction of the mechanical properties of the molded component [1–3]. One cause of the formation of voids in fiber composite components is inhomogeneous flow processes at microscopic levels inside the textile preforms. Depending on the process parameters, the resin system flows at different rates within the tows of the reinforcement fabrics and channels between the tows, due to the different permeability of the two areas. As shown in Figure 1, a flow front is formed, which can be divided into a saturated, partially saturated, and unsaturated region of the tows [4,5].

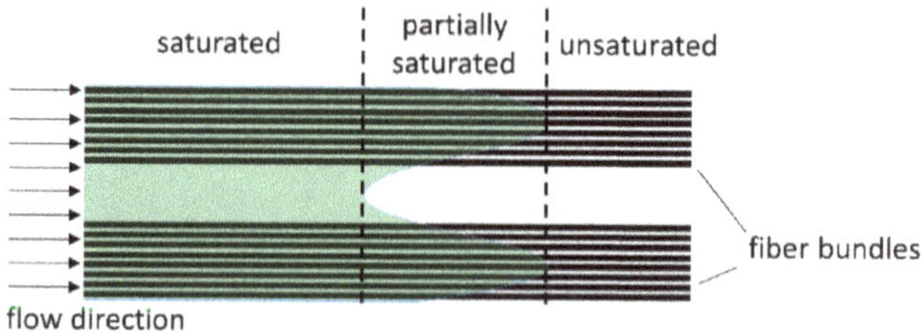

Figure 1. Saturated, partially saturated, and unsaturated regions of the flow front in a unidirectional preform.

This effect, known as dual-scale flow, is the focus of several studies since it is a major reason for the formation and transport of voids [6–11]. If the flow velocities within the tows and channels between the tows match, void-free components are produced [12–15] and the lightweight potential of the materials is optimally exploited.

The flow of the liquid resin system through the textile preform as a porous medium is described by Darcy's law (Equation (1)).

$$v_m = -\frac{\vec{K}}{\eta} \cdot \nabla p \tag{1}$$

where v_m is the volume-averaged velocity, $\vec{K}$ the permeability tensor, η the resin viscosity, and ∇p the pressure gradient from inlet to the flow front position.

Assuming that the resin system is an incompressible medium, the law of conservation of mass applies:

$$\nabla \cdot v_m = 0 \tag{2}$$

The semi-empirical Kozeny-Carman equation can be used to determine the macroscopic permeability K of the textile preform [16]:

$$K = \frac{r_f^2}{4k_c} \frac{\left(1 - \varphi_f\right)^3}{\varphi_f{}^2} \tag{3}$$

where r_f is the fiber radius, φ_f is the fiber volume fraction, and k_c is the Kozeny constant. The Kozeny constant is highly dependent on the resin used, impregnation direction, and textile preform [17], and thus is not precisely determined [9,18]. Nevertheless, this model allows calculations on the progression of the flow front in the textile preform and the resulting process duration.

However, this model is not suitable for a more detailed consideration of the impregnation of textile preforms [9] since no information is obtained about the microscopic flow processes within and between the tows. The proportion of resin flowing into the individual fiber bundles of a textile preform during impregnation is described in the numerical simulations by means of an extension by a loss term q, which depends on pressure p and degree of saturation s [6]:

$$\nabla \cdot \left(\frac{\vec{K}}{\eta} \cdot \nabla p \right) = q(p, s) \tag{4}$$

Analytically, the phenomenon of dual-scale flow can be represented by the modified capillary number Ca^*. The modified capillary number forms the ratio of viscosity-dependent and capillary force-dependent flow [8,13] and thus allows conclusions about the proportions of the unsaturated region of the flow front [14].

$$Ca^* = \frac{\mu \cdot \overline{u}}{\gamma \cdot \cos\theta} \tag{5}$$

where Ca^* is the modified capillary number, μ the dynamic resin viscosity, $\overline{u}$ the averaged macroscopic flow velocity, γ the surface tension, and θ the contact angle between the resin and fibers.

At high injection pressures, the proportion of viscosity-dependent flow predominates. The channels between the tows fill faster than the areas within the tows. This results in the inclusion of elongated micropores in the tows (Figure 2b). If capillary forces prevail, the flow front within the tows progresses faster. Spherical macropores are formed in the channels (Figure 2a).

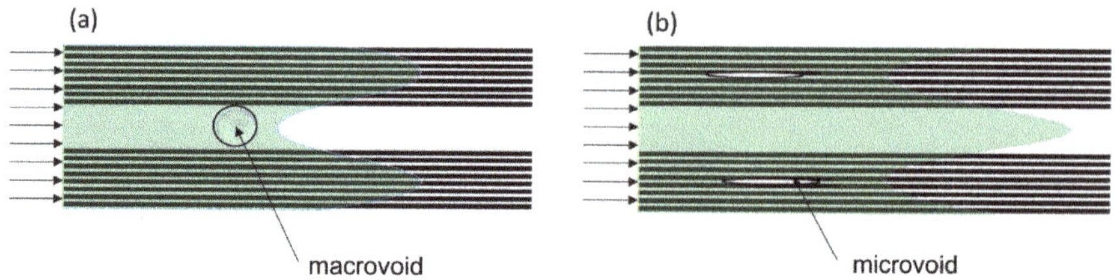

Figure 2. Formation of voids in the dual-scale model; (**a**) Formation of spherical macrovoids in the channels between the tows; (**b**) Formation of elongated microvoids inside the tows.

With the help of the modified capillary number during injection, in conjunction with the geometric structure of the textile preforms, conclusions can be drawn about the void formation in the component [12–15,19].

Models for calculating the resulting void volume content are based on the ratio of the flow front progress within and between the tows. Gueroult et al. [15] contrast the two time scales of the flow time inside the fiber bundles Δt_t in relation to the flow time in the channels Δt_c.

$$\frac{\Delta t_t}{\Delta t_c} = \frac{K_c}{K_t} \cdot (1 - \varphi_{FT}) \cdot \left[1 - \frac{F_S \cdot K_c \cdot \varphi_{FT}}{d_{Fi} \cdot (1 - \varphi_{FT}) \cdot L_t \cdot Ca^*} \cdot ln\left(\frac{Ca^* \cdot d_{Fi} \cdot (1 - \varphi_{FT}) \cdot L_t)}{F_S \cdot K_c \cdot \varphi_{FT}} + 1\right)\right] \quad (6)$$

where K_c is the permeability of the channel, K_t the permeability of the tow, φ_{FT} the fiber volume content in the tow, F_S a shape factor depending on longitudinal or transversal flow direction, d_{Fi} the diameter of a single fiber, and L_t the length of a tow.

A ratio of $\frac{\Delta t_t}{\Delta t_c} < 1$ describes the advance of the resin within the fiber bundles and resulting emergence of macropores. $\frac{\Delta t_t}{\Delta t_c} > 1$ implies a faster advance of resin within the channels and the formation of microvoids. At a ratio of $\frac{\Delta t_t}{\Delta t_c} = 1$, the resin flows at identical velocities in both sections, which means that no air can be entrapped, and no voids are formed due to the dual-scale flow [15].

Validation of such "dual-scale" computational models usually involves evaluating the resulting void volume contents and classifying them into microvoids and macrovoids [20,21]. An alternative is the optical detection of the different flow velocities. The dual-scale effect was investigated by several studies [10,22,23] using a microscope locally during injection. The challenge in this type of analysis is to find a compromise between maximizing the image section of the flow front while retaining locally high resolution. Another possibility to prove the phenomenon of dual-scale flow is numerical simulation. Godbole et al. [24] describe the differences in the flow velocity within and between the fiber bundles by determining the length of the partially saturated flow front L_{ps}, by simulations (Figure 3):

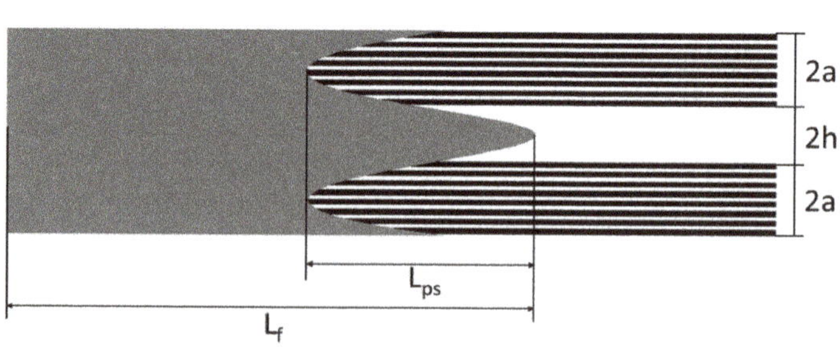

Figure 3. Length of the partially saturated flow front L_{ps}.

Neglecting capillary forces, it is shown that the length of the partially saturated zone remains constant if the flow path is sufficiently long. This finding agrees with the results of Zhou et al. [25,26], who also calculate a constant length of the partially saturated zone.

The resulting length depends on the permeability of the textile preforms and the preform geometry, as well as the volume fraction of the fiber bundles of a unidirectional (UD) unit cell:

$$L_{ps} = \sqrt{\frac{2a}{K_{yytow}}\left[\frac{h^3}{3}\right]} \cdot V_{tow-ply} \quad (7)$$

where a is half of the width of a tow, K_{yytow} the transverse permeability of a tow, and h half of the width of the channel between the tows. $V_{tow-ply}$ is the volume fraction of tows inside a UD unit cell and calculated from the proportion of fiber bundles as closed solids in the cross-section of the laminate:

$$V_{tow-ply} = \frac{a}{h+a} \quad (8)$$

Detailed information about the microscopic dual-scale flow and associated pore formation becomes possible when the state of the impregnation can be imaged holistically in a cavity. The state of the art in investigating void formation and transport are optical methods [27–29]. Furthermore, ultrasonic measurements, as well as X-ray and micro-CT examinations, were conducted in several renowned studies [1].

However, all the mentioned techniques are currently not suited to analyze big areas of the partially saturated zone, because of their limited resolution or the requirement of additives to increase visibility. Due to the limited field of view, a "snapshot" approach to optical measurement of the void distribution is needed; however, a suitable method has not yet been presented for this issue [30].

The novel approach presented in this article is to freeze the complete impregnation process at different filling levels to investigate the flow front section by section. To obtain snapshots of the component impregnation, a methodology is developed below that uses spontaneous curing of a photopolymerizing resin. Unique to this method is the gathering of specimen with spontaneously cured partially saturated flow fronts, which can be holistically observed via microscopy.

The resin systems used for this purpose include photoreactive functional groups that crosslink when exposed to light [31]. Components of such resin systems are monomers, oligomers, and photo initiators. Upon absorption of high-energy light, mostly in the ultraviolet spectrum, the photo initiators form radicals or ions. These serve as initiators for the crosslinking reaction between oligomers and monomers [32,33].

Depending on the resin system used and film thickness, the time required for complete crosslinking can range from a fraction of a second to several minutes [32]. To minimize interferences caused by changing pressure gradients during crosslinking, it is advantageous to react as quickly as possible. Only in the case of spontaneous crosslinking of the complete molded part are the pore formation as well as the pore transport frozen in situ.

Based on the frozen filling samples produced with photopolymerizing resins, dual-scale flow in fiber bundles and channels is investigated sequentially along the complete flow front by means of microscopy. The images obtained are used to compare current computational models with the optical measurements. The evaluation of the flow conditions can be used to verify new calculation models and increase the accuracy of FEM calculations.

2. Materials and Methods

2.1. Experimental Setup

An injection mold for linear impregnation of textile preforms was designed. It consists of a bottom side made of aluminum with a linear gate and riser, on which a single-layer textile preform of 300 mm length and 130 mm width is draped. The opposite side of the mold consists of a glass plate fixed with an aluminum frame (Figure 4). By means of defined torque for the screws positioned circumferentially on the frame, a constant compression pressure of 4.4 MPa is set.

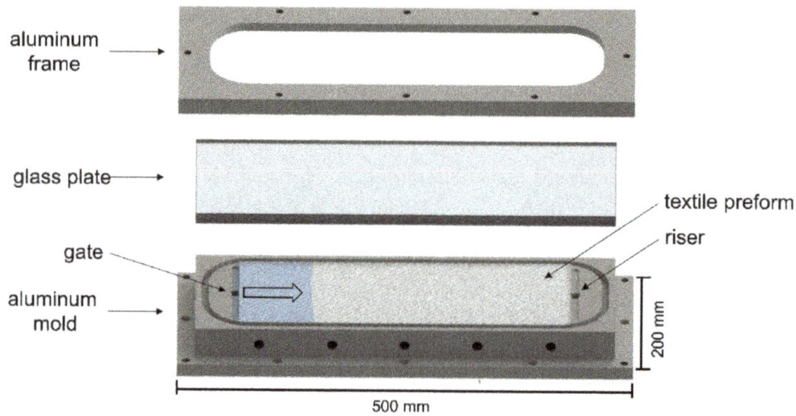

Figure 4. Experimental setup to produce planar test specimens with a photoreactive resin system.

The glass plate is transparent to light in the ultraviolet (UV) range. UV spotlights are mounted above the mold, which completely and uniformly illuminate the cavity when switched on. A photoreactive resin system is used which, when irradiated with UV light, stops the flow process without any noticeable delay. Three YG-TGD20-405 LED emitters from Shenzhen Creality 3D Technology Co, Ltd., Shenzhen, China, with an emitted waveband of 400 nm to 405 nm, at an overall system power of 7.8 W each, are used at 100% intensity. The resin system is injected under constant injection pressure in various gradations.

The mold is divided into five sections and the flow front is stopped after every 60 mm by switching on the UV lamps. This is followed by an exposure time of 60 s, during which the resin system cures completely.

This methodology allows the examination of the entire flow front of the fabricated specimens after curing. Optical studies were performed using a Keyence VHX-7000 microscope at 100× magnification. The aim of the measurements is a quantified mapping of the dual-scale flow behavior for comparison with the calculation results of the theoretical models. For this purpose, the flow front progress within the tows l_t and in the channels between the tows l_c, as well as their flow path difference Δl, is measured (see Figure 5).

The process of impregnation is recorded by a camera to measure the mean flow velocity within the sections and calculate the modified capillary number according to Equation (5).

The resin system is injected with constant injection pressure. The examined parameter combinations are shown in Figure 6. Each combination is repeated three times so that a total of 60 test specimens is produced and evaluated. The last mold section with a flow path length of 300 mm is excluded, since the textile preforms could not be completely saturated over the entire length at pressure levels of 0.05 MPa and 0.1 MPa. The pressure levels were selected to show clearly distinct flow path differences; however, the pronounced dual-scale flow leads to a very high number of overlapping microvoids that cannot be thoroughly evaluated in this study.

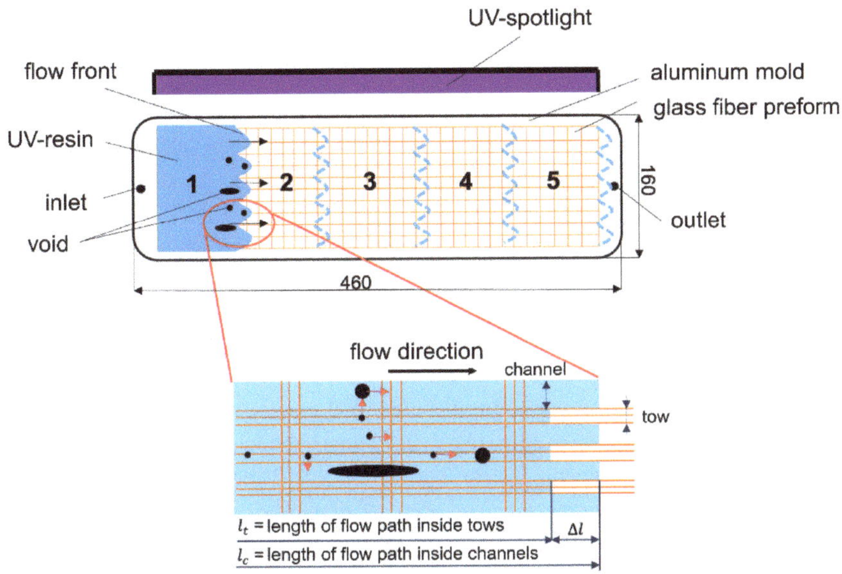

Figure 5. Schematic course of the flow front and division of the flow path into mold sections denominated from inlet to outlet as one to five.

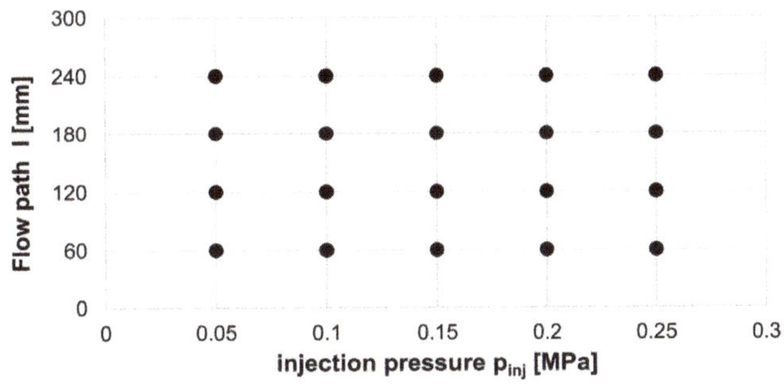

Figure 6. Parameter combinations for injection experiments.

For each parameter combination 20 flow path lengths in the fiber bundle l_t and between the fiber bundles l_c are determined, their ratio is calculated, and the values are compared with the models of Gueroult [15] (Equation (6)) and Godbole [24] (Equation (7)).

2.2. Materials

The resin system used is a photoreactive 3D Printing UV sensitive resin from the manufacturer Shenzhen Anycubic Technology Co., Ltd., Shenzhen, China. It is composed of 30% to 60% oligomers (polyurethane acrylate), 10% to 40% acrylate monomers and 2% to 5% photo initiator. All experiments were conducted with the same batch of the resin system to exclude variations in the components between the experiments.

The textile semi-finished product used is a glass fiber filament fabric type 92130 from Porcher Industries Germany GmbH, Erbach, Germany for which the following data was measured, as depicted in Table 1. A fabric with medium grammage and plain weave with a high number of crossovers was chosen to establish a homogeneous flow while producing samples with conventional thickness. Moreover, the fabric is resilient against fiber displacement in the manual preparation process, which can potentially cause local deviations in permeability.

Table 1. Measured and calculated parameters of the fabric.

Parameter	Abbreviation	Value	Unit
Type of weave		Plain weave	[-]
Grammage	m_f	395	[g/m^2]
Thread count warp		6	[L/cm]
Thread count weft		6.5	[L/cm]
Fiber diameter	d_f	9	[µm]
Width of tows	$2a$	1.38	[mm]
Width of channels	$2h$	0.39	[mm]
Transverse permeability of tows	K_{yytow}	5.32×10^{-7}	[mm^2]
Volume fraction of tows in a unit cell	$V_{tow-ply}$	77.99	[%]

3. Results and Discussion

3.1. Experimental Results

During the entire injection period of all specimens, the uniform, approximately linear progress of the flow front must be ensured in order to avoid volume flows in the transverse direction and comply with the boundary conditions of the continuity equation according to Darcy (Equation (1)). Specimens with irregular flow fronts are excluded from the evaluation and prepared again. Several individual images of the flow front are taken with a 100× magnification factor and assembled to form a complete image of the frozen flow front, Figure 7a,b.

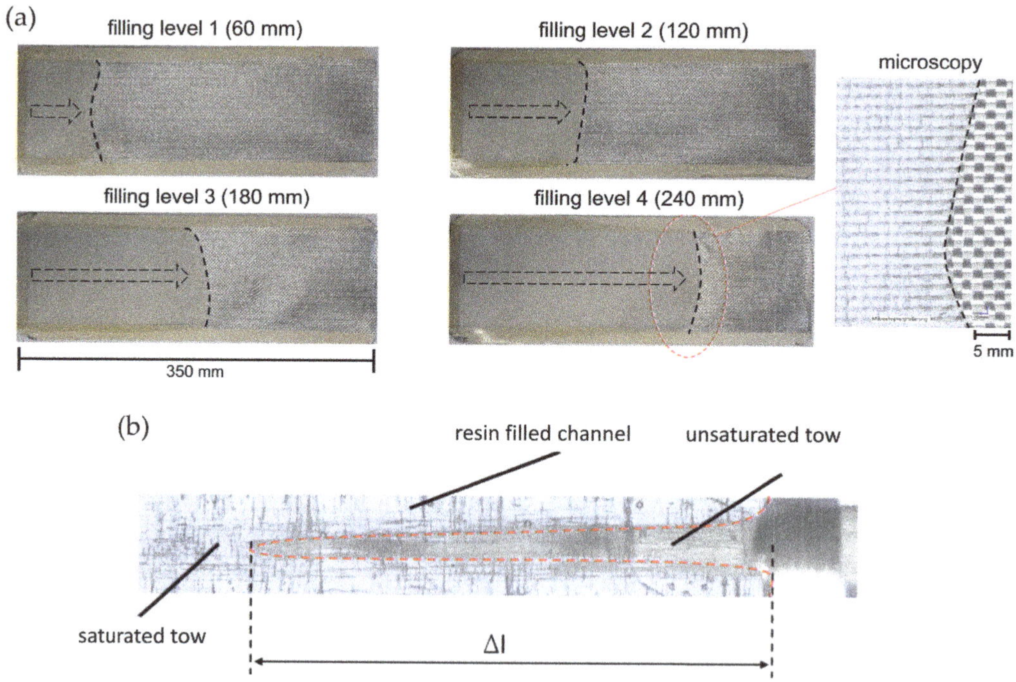

Figure 7. (**a**) Frozen flow front in different mold sections and merged microscopy image; (**b**) single flow path difference Δl, shown magnified.

The frozen flow front of the cured specimen shows a distinct edge between the impregnated and dry sections. These definite lines indicate a virtually instant pervasive curing throughout the complete thickness of the transparent composite material and exist at each applied pressure gradient.

As in the literature [24–26], there are no distinct changes in the flow path difference along the mold cavity at constant process parameters, as shown in the comparison in Figure 8. However, there is a clear dependence of the flow path difference on the injection pressure.

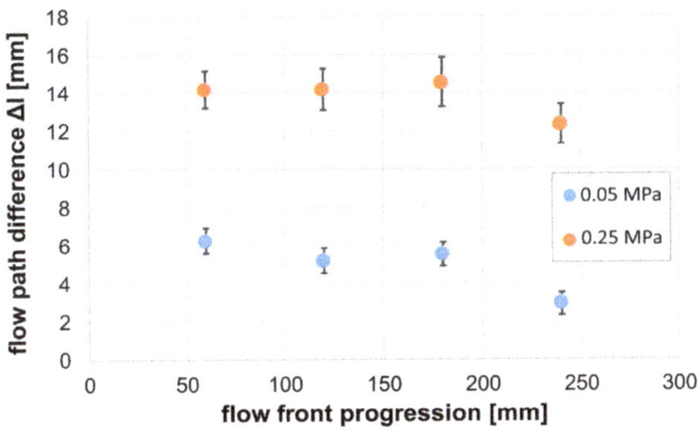

Figure 8. Flow path difference along the mold sections.

The existing scatter is mainly attributed to fluctuations in permeability caused by the manual insertion of the fabrics into the cavity, since the manual preparation of the preforms may result in local displacements of the tows, and thusly produce irregular gaps and fiber distributions.

The injection pressure has a considerable influence on the flow velocity and thus on the locally prevailing capillary number. Contrary to what was calculated by Godbole [24], the results averaged per experiment show the proportionality of the flow path difference and the capillary number to the injection pressure (Figure 9). This measurable effect supports the model of Gueroult [15], in which the flow time ratio is largely determined by the capillary number. According to Equations (1) and (5), the capillary number is also pressure dependent. The calculations of Godbole [24] imply that a stronger expression of the dual-scale flow is completely compensated for by increased crossflow effects. This assumption is not confirmed by the measurement of the flow path differences for the presented experimental setup.

Considering all local measurements of the flow path difference as a function of the modified capillary number, an increase in flow path difference with an increasing modified capillary number is observed (Figure 10). At capillary numbers above around 0.025, the increase in the flow path difference is less substantial. The flattened slope is in accordance with the natural logarithm of Equation (6), which determines the flow ratio inside tows and channels. Additional experiments must be conducted to verify the shape of the curve for higher capillary numbers.

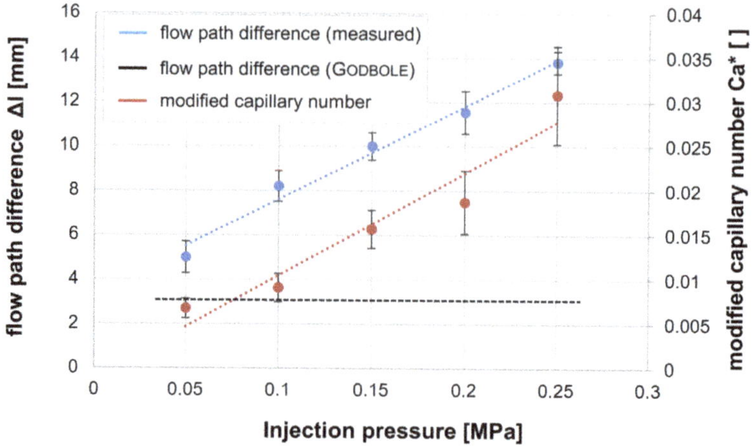

Figure 9. Dependence of the flow path difference and modified capillary number on the injection pressure.

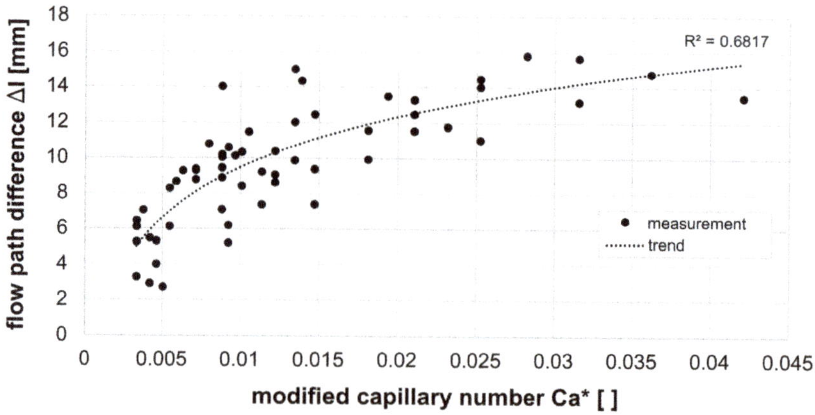

Figure 10. Trend of the flow path difference in coherence with the local modified capillary number.

The increase in the flow path difference with increasing capillary number supports the findings of renowned studies [12–14,19] for process conditions with predominantly pressure-induced flow in the dual-scale model. Each specimen is in the region of micropore formation ($\frac{\Delta t_t}{\Delta t_c} > 1$). With increasing injection pressure, this flow time ratio also increases.

The measurement results illustrate that this increase is also reflected in increasing flow path differences. The progression of the flow path difference towards a maximum value implies an increase in crossflow effects, which counteracts the further increase in the flow path difference.

The measurements show that the flow velocity in the fiber bundles is slower than in the channels between the bundles. Neglecting crossflow effects, the flow path difference can be inferred from the flow time ratio according to the model of Gueroult [15].

$$\Delta l = \overline{v} \cdot \Delta t \tag{9}$$

Since the flow state in the bundle and channel is stopped after the identical injection time Δt when the UV illumination is activated, it follows:

$$\Delta l_c(t) = \overline{v_c} \cdot \Delta t \tag{10}$$

$$\Delta l_t(t) = \overline{v_t} \cdot \Delta t \tag{11}$$

$$\frac{\Delta l_c(t)}{\Delta l_t(t)} = \frac{\overline{v_c}}{\overline{v_t}} \cdot \Delta t = c_v \cdot \Delta t \tag{12}$$

However, Gueroult's model [15] is based on the ratio of the flow times of the resin system to saturation of the length Δl of a single unit cell:

$$\Delta t_c(l) = \frac{\Delta l}{\overline{v_c}} \tag{13}$$

$$\Delta t_t(l) = \frac{\Delta l}{\overline{v_t}} \tag{14}$$

$$\frac{\Delta t_c(l)}{\Delta t_t(l)} = \frac{\overline{v_t}}{\overline{v_c}} \cdot \Delta l = \frac{1}{c_v} \cdot \Delta l \tag{15}$$

Equations (12) and (15) indicate inverse proportionality of length ratios and time ratios of impregnation of fiber bundles and channels.

$$\frac{\Delta l_c(t)}{\Delta l_t(t)} \sim \frac{\Delta t_t(l)}{\Delta t_c(l)} \tag{16}$$

If the ratios are plotted on top of each other as a function of the injection pressure, a good agreement of the values is observed for the first mold section, see Figure 11.

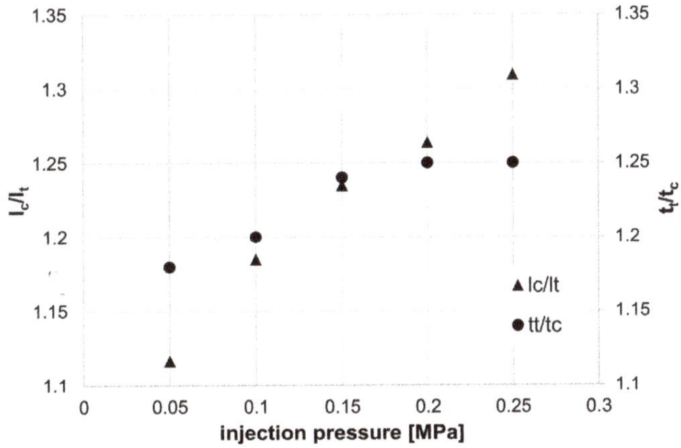

Figure 11. Comparison of the measured length ratios at 60 mm distance from the sprue with the calculated time ratios according to Equation (6).

Due to crossflow of the resin out of the channels into the fiber bundles in the transverse direction, as shown in Figure 12, the flow path difference Δl stagnates while the macroscopic flow path continues to increase.

As the flow path length progresses, the results of the successive mold sections (Figure 5) therefore show considerable deviations (Figure 13).

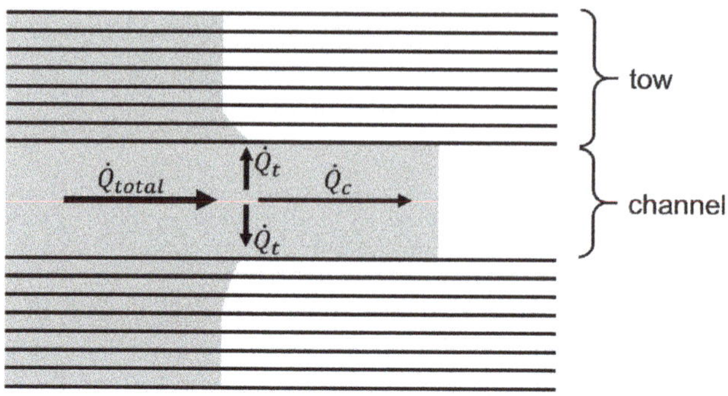

Figure 12. Division of the total volume flow into crossflow $\dot{Q}_t$ into the tow and longitudinal flow $\dot{Q}_c$ in the channel.

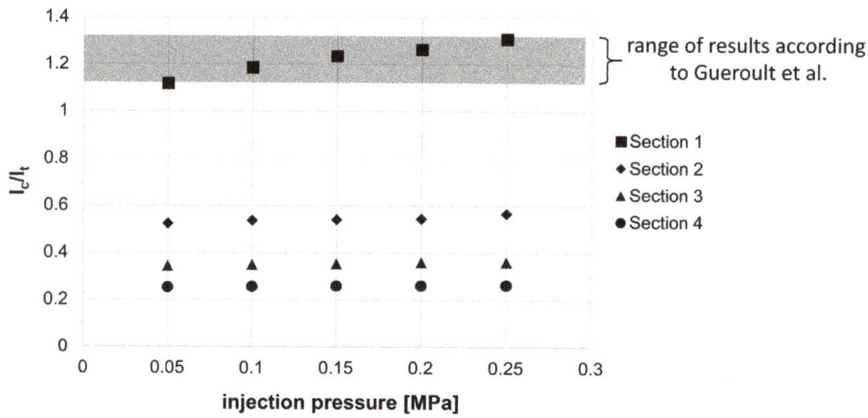

Figure 13. Comparison of measured aspect ratios of all tool sections.

The deviations of the flow path ratios from the flow time ratios can be adjusted by correction factors. As Bodaghi et al. [9] describe, a loss term $q\ (p, s)$ from Equation (4) contributes to the transverse impregnation of tows. The regression analysis of the measurement results confirms the assumption that both the injection pressure and degree of saturation of the macroscopic flow path influence the loss term, as shown in the comparison in Figure 14 While the slopes of the compensation lines are determined by the injection pressure, the intercept is related to the macroscopic flow path and is thus a measure of the degree of saturation of the textile preform. It follows:

$$\frac{\Delta t_t(l)}{\Delta t_c(l)} = c_{corr} \cdot \frac{\Delta l_c(t)}{\Delta l_t(t)} \tag{17}$$

$$c_{corr} = a \cdot p_{inj} + b(s) \tag{18}$$

where c_{corr} is a correction factor, a is the slope factor, p_{inj} the injection pressure, and $b(s)$ the saturation dependent offset.

This correction factor c_{korr} and its dependence on pressure and saturation proves the occurrence of the loss term $q\ (p, s)$ described by Bodaghi et al. and quantifies the effect of crossflow for the given material and process parameter combination.

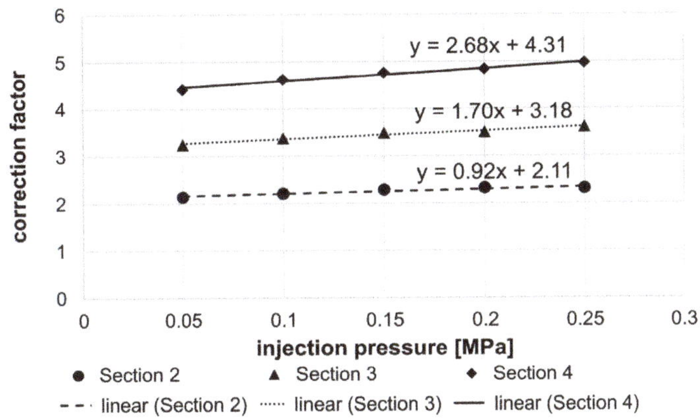

Figure 14. Determined correction factors for the adjustment of the measured length ratios to the calculated time ratios according to Gueroult.

3.2. Discussion

Two distinct aspects must be considered when analyzing the results. First is the applicability of the novel methodology, followed by the observations regarding dual-scale flow experiments.

The presented methodology allows taking snapshots of the flow processes during the impregnation of textile preforms using a photoreactive resin system. Evidence was provided that photoreactive resin systems crosslink sufficiently fast to freeze the microscopic saturation along the entire specimen. The curing occurred sufficiently rapidly for the given glass fiber fabric and applied pressure settings. Upper limitations of flow velocity and specimen thickness must be determined in prospective studies to eliminate the possibility of partial flow progression below the irradiated surface. In contrast to point-wise optical measurements, the complete unsaturated domain of the flow front can be analyzed. The holistic analysis of the flow front represents an improvement compared with the limited field of view of well-established methods [27,28,30]. It is more cost- and time-efficient than x-ray or micro-CT measurements [1]. Depending on the type of microscope, voids with diameters of 5 to 20 μm can be analyzed [1], whereas ultrasonic imaging is only suitable for single defects with a minimum size between 1 and 0.6 mm [34,35]. The simple mold setup and its similarity to live-microscopy makes the introduced method easily accessible. The glass cover allows the elimination of race tracking inside the entire mold during impregnation, which cannot be ruled out with single point-wise observations, as Siddig et al. [36] proved, at least three points of a rectangular shape need to be observed to detect race tracking. However, one downside of this method is that it is not suited to quantify void transport. Freezing the flow produces a single image that can be used to describe the dual flow effect and formation of voids at the very front of the flow, but it contains no information on the history of voids inside the specimen.

The result of the experimental setup provides a multitude of insights into the saturation process of fabrics. A clear dependence on the injection pressure was determined for the measured flow path difference. This effect can be explained by Equations (4)–(6), which show that the flow time ratios of Gueroult [15] depend on the flow velocity caused by the injection pressure gradient. A comparison of the results of the established relationship of the flow paths in the fiber bundle and channels between the bundles with the model of the flow time ratios of Gueroult [15] shows good agreement within the first mold section with a 60 mm flow path. With an increasing flow path, the deviations of the time and path ratios become larger, which is due to the increasing influence of crossflows in the transverse fiber direction. These crossflows are summarized by Bodaghi et al. [9] in a loss term, which depends on the injection pressure as well as the degree of saturation. The results of the measurements show that such a loss term can be described by a correction factor that was determined for each section of the mold and the corresponding injection pressure Simulations of previous studies [24–26] showed that using constant tool and material parameters, the flow front difference remains unchanged with sufficient tool length. These results could only be partially confirmed. While the flow front difference remains constant along the different mold sections within the individual parameter combinations, in contrast to the studies of Godbole [24] and Zhou [25,26], the influence of injection pressure cannot be neglected. However, there are considerable differences in the experimental setup and simulation constraints. While the simulations [24–26] use perfectly unidirectional tows, the experimental setup was conducted with plain weave fabric and a different type of inlet. Furthermore, the simulated tow permeability K_{yytow} was higher than the determined permeability of the experiments. Both factors result in differences in longitudinal and transverse flow and have an influence on the impact of capillary forces. Further experiments should be conducted that replicate the constraints of the simulations of Godbole and Zhou to be able to compare results.

4. Conclusions

With the presented methodology, the flow front of the cross-linked specimens can be viewed completely and contiguously at high resolution, which is a significant improvement over locally confined in-situ measurements. The measurement methodology represents a good starting point for the creation and validation of calculation models for the impregnation of textile preforms. Optimization potential lies in the tool design and manual preparation of the textile preforms. The adjustment of the compression pressure of the textile preforms by means of a defined tightening torque of the screw connection is reproducible to a limited extent and leads to fluctuating permeability. This deviation can be eliminated by using a vertically loosely supported plate as the mold surface with pressure sensors underneath. To specify exact physically based models for the calculation of the correction factors, either a more precise adjustability of the compression pressure must be ensured, or the tow permeability must be evaluated individually for each experiment by cutting the cured specimen and determining the surface ratios of fiber bundles and resin of the cross section. Analytical models of the dual-scale flow were confirmed and an approach to quantify the loss term of transverse flow is given. The prevalent increase of flow path differences differs from simulations, which emphasizes the need of further investigations. Since only a limited number of experiments were conducted, additional experiments should be performed to validate the findings. In future experiments, a variety of textile preforms must be analyzed to verify the applicability of the method for diverse types of weaves. Furthermore, higher modified capillary numbers must be examined, by increasing the macroscopic resin velocity via increased injection pressure. After validation of the methodology, future experiments can be conducted to directly evaluate resulting void volume contents and the coherence with the now measurable flow path differences.

Author Contributions: Conceptualization, B.N.; methodology, B.N. and F.P.; validation, B.N.; formal analysis, F.P.; investigation, B.N.; resources, F.P.; data curation, B.N.; writing—original draft preparation, B.N.; writing—review and editing, F.P.; visualization, B.N.; supervision, F.P.; project administration, F.P. All authors have read and agreed to the published version of the manuscript.

Funding: This research received no external funding.

Institutional Review Board Statement: Not applicable.

Informed Consent Statement: Not applicable.

Acknowledgments: We acknowledge support for the publication costs from the Open Access Publication Fund of the Technische Universität Ilmenau and the infrastructural support of the Thüringer Innovationszentrum Mobilität (ThIMo).

Conflicts of Interest: The authors declare no conflict of interest.

References

1. Mehdikhani, M.; Gorbatikh, L.; Verpoest, I.; Lomov, S.V. Voids in fiber-reinforced polymer composites: A review on their formation, characteristics, and effects on mechanical performance. *J. Compos. Mater.* **2019**, *53*, 1579–1669. [CrossRef]
2. Hagstrand, P.-O.; Bonjour, F.; Månson, J.-A.E. The influence of void content on the structural flexural performance of unidirectional glass fibre reinforced polypropylene composites. *Compos. Part A Appl. Sci. Manuf.* **2005**, *36*, 705–714. [CrossRef]
3. De Almeida, S.F.M.; Neto, Z.d.S.N. Effect of void content on the strength of composite laminates. *Compos. Struct.* **1994**, *28*, 139–148. [CrossRef]
4. Kuentzer, N.; Simacek, P.; Advani, S.G.; Walsh, S. Permeability characterization of dual scale fibrous porous media. *Compos. Part A Appl. Sci. Manuf.* **2006**, *37*, 2057–2068. [CrossRef]
5. Binétruy, C.; Hilaire, B.; Pabiot, J. The interactions between flows occurring inside and outside fabric tows during rtm. *Compos. Sci. Technol.* **1997**, *57*, 587–596. [CrossRef]
6. Facciotto, S.; Simacek, P.; Advani, S.G.; Middendorf, P. Modeling of anisotropic dual scale flow in RTM using the finite elements method. *Compos. Part B Eng.* **2021**, *214*, 108735. [CrossRef]
7. Carlone, P.; Rubino, F.; Paradiso, V.; Tucci, F. Multi-scale modeling and online monitoring of resin flow through dual-scale textiles in liquid composite molding processes. *Int. J. Adv. Manuf. Technol.* **2018**, *96*, 2215–2230. [CrossRef]
8. Batyrshin, E.S.; Solnyshkina, O.A.; Pityuk, Y.A. Study of the Features of Double Porosity Media Impregnation. *Technol. Phys.* **2022**, *24*, 852. [CrossRef]
9. Bodaghi, M.; Lomov, S.V.; Simacek, P.; Correia, N.C.; Advani, S.G. On the variability of permeability induced by reinforcement distortions and dual scale flow in liquid composite moulding: A review. *Compos. Part A Appl. Sci. Manuf.* **2019**, *120*, 188–210. [CrossRef]
10. Yashiro, S.; Nakashima, D.; Oya, Y.; Okabe, T.; Matsuzaki, R. Particle simulation of dual-scale flow in resin transfer molding for process analysis. *Compos. Part A Appl. Sci. Manuf.* **2019**, *121*, 283–288. [CrossRef]
11. Schell, J.S.U.; Deleglise, M.; Binetruy, C.; Krawczak, P.; Ermanni, P. Numerical prediction and experimental characterisation of meso-scale-voids in liquid composite moulding. *Compos. Part A Appl. Sci. Manuf.* **2007**, *38*, 2460–2470. [CrossRef]
12. Ruiz, E.; Achim, V.; Soukane, S.; Trochu, F.; Breard, J. Optimization of injection flow rate to minimize micro/macro-voids formation in resin transfer molded composites. *Compos. Sci. Technol.* **2006**, *66*, 475–486. [CrossRef]
13. Park, C.H.; Lebel, A.; Saouab, A.; Bréard, J.; Lee, W.I. Modeling and simulation of voids and saturation in liquid composite molding processes. *Compos. Part A Appl. Sci. Manuf.* **2011**, *42*, 658–668. [CrossRef]
14. Teixidó, H.; Staal, J.; Caglar, B.; Michaud, V. Capillary Effects in Fiber Reinforced Polymer Composite Processing: A Review. *Front. Mater.* **2022**, *9*, 37. [CrossRef]
15. Gueroult, S.; Lebel-Lavacry, A.; Park, C.H.; Bizet, L.; Saouab, A.; Bréard, J. Analytical modeling and in situ measurement of void formation in liquid composite molding processes. *Adv. Compos. Mater.* **2013**, *23*, 31–42. [CrossRef]
16. Carman, P.C. Fluid flow through granular beds. *Chem. Eng. Res. Des.* **1997**, *75*, S32–S48. [CrossRef]
17. Amico, S.; Lekakou, C. An experimental study of the permeability and capillary pressure in resin-transfer moulding. *Compos. Sci. Technol.* **2001**, *61*, 1945–1959. [CrossRef]
18. Gutowski, T.G.; Morigaki, T.; Cai, Z. The Consolidation of Laminate Composites. *J. Compos. Mater.* **1987**, *21*, 172–188. [CrossRef]
19. Lee, G.-W.; Lee, K.-J. Mechanism of Void Formation in Composite Processing with Woven Fabrics. *Polym. Polym. Compos.* **2003**, *11*, 563–572. [CrossRef]
20. Oya, Y.; Matsumiya, T.; Ito, A.; Matsuzaki, R.; Okabe, T. Gate optimization for resin transfer molding in dual-scale porous media: Numerical simulation and experiment measurement. *J. Compos. Mater.* **2020**, *54*, 2131–2145. [CrossRef]
21. Leclerc, J.S.; Ruiz, E. Porosity reduction using optimized flow velocity in Resin Transfer Molding. *Compos. Part A Appl. Sci. Manuf.* **2008**, *39*, 1859–1868. [CrossRef]
22. LeBel, F.; Fanaei, A.E.; Ruiz, É.; Trochu, F. Prediction of optimal flow front velocity to minimize void formation in dual scale fibrous reinforcements. *Int. J. Mater. Form.* **2014**, *7*, 93–116. [CrossRef]
23. Imbert, M.; Comas-Cardona, S.; Abisset-Chavanne, E.; Prono, D. Experimental investigation of intra-tow fluid storage mechanisms in dual-scale fiber reinforcements. *Compos. Part A Appl. Sci. Manuf.* **2018**, *107*, 70–82. [CrossRef]
24. Godbole, M.G.; Gururaja, S.; Joshi, M.; Advani, S. Semi-analytical formulation of effective permeability of a dual scale unidirectional fabric. *Compos. Part A Appl. Sci. Manuf.* **2021**, *150*, 106630. [CrossRef]
25. Zhou, F.; Kuentzer, N.; Simacek, P.; Advani, S.G.; Walsh, S. Analytic characterization of the permeability of dual-scale fibrous porous media. *Compos. Sci. Technol.* **2006**, *66*, 2795–2803. [CrossRef]
26. Zhou, F.; Alms, J.; Advani, S.G. A closed form solution for flow in dual scale fibrous porous media under constant injection pressure conditions. *Compos. Sci. Technol.* **2008**, *68*, 699–708. [CrossRef]
27. Patel, N.; Rohatgi, V.; Lee, L.J. Micro scale flow behavior and void formation mechanism during impregnation through a unidirectional stitched fiberglass mat. *Polym. Eng. Sci.* **1995**, *35*, 837–851. [CrossRef]
28. Chen, Y.-T.; Davis, H.T.; Macosko, C.W. Wetting of fiber mats for composites manufacturing: I. Visualization experiments. *AIChE J.* **1995**, *41*, 2261–2273. [CrossRef]
29. Lystrup, C.; George, A.; Zobell, B.; Boster, K.; Childs, C.; Girod, H.; Fullwood, D. Optical measurement of voids in situ during infusion of carbon reinforcements. *J. Compos. Mater.* **2021**, *55*, 775–786. [CrossRef]

30. Pillai, K.M. Modeling the Unsaturated Flow in Liquid Composite Molding Processes: A Review and Some Thoughts. *J. Compos. Mater.* **2004**, *38*, 2097–2118. [CrossRef]
31. Crivello, J.V.; Reichmanis, E. Photopolymer Materials and Processes for Advanced Technologies. *Chem. Mater.* **2014**, *26*, 533–548. [CrossRef]
32. Ravve, A. *Light-Associated Reactions of Synthetic Polymers*; Springer Science + Business Media LLC.: New York, NY, USA, 2006; ISBN 0-387-31803-8.
33. Pagac, M.; Hajnys, J.; Ma, Q.-P.; Jancar, L.; Jansa, J.; Stefek, P.; Mesicek, J. A Review of Vat Photopolymerization Technology: Materials, Applications, Challenges, and Future Trends of 3D Printing. *Polymers* **2021**, *13*, 598. [CrossRef] [PubMed]
34. Tian, F.; Hao, Y.; Zou, Z.; Zheng, Y.; He, W.; Yang, L.; Li, L. An Ultrasonic Pulse-Echo Method to Detect Internal Defects in Epoxy Composite Insulation. *Energies* **2019**, *12*, 4804. [CrossRef]
35. Taheri, H.; Hassen, A.A. Nondestructive Ultrasonic Inspection of Composite Materials: A Comparative Advantage of Phased Array Ultrasonic. *Appl. Sci.* **2019**, *9*, 1628. [CrossRef]
36. Siddig, N.A.; Binetruy, C.; Syerko, E.; Simacek, P.; Advani, S. A new methodology for race-tracking detection and criticality in resin transfer molding process using pressure sensors. *J. Compos. Mater.* **2018**, *52*, 4087–4103. [CrossRef]

Journal of
Composites Science

MDPI

Article

Numerical Simulation of Two-Phase Flow in Liquid Composite Moulding Using VOF-Based Implicit Time-Stepping Scheme

Hatim Alotaibi [1], Chamil Abeykoon [2,3], Constantinos Soutis [2,3] and Masoud Jabbari [1,*]

[1] Department of Mechanical, Aerospace and Civil Engineering, The University of Manchester, Manchester M13 9PL, UK
[2] Department of Materials, The University of Manchester, Manchester M13 9PL, UK
[3] Aerospace Research Institute, The University of Manchester, Manchester M13 9PL, UK
* Correspondence: m.jabbari@manchester.ac.uk

Abstract: The filling stage in injection/infusion moulding processes plays a key role in composite manufacturing that can be influenced by the inlet and vent ports. This will affect the production of void-free parts and the desirable process time. Flow control is usually required in experiments to optimise such a stage; however, numerical simulations can be alternatively used to predict manufacturing-induced deficiencies and potentially remove them in the actual experiments. This study uses ANSYS Fluent software to model flow-front advancement during the impregnation of woven fabrics. A developed technique is applied by creating tracking points (e.g., on-line monitor) in the direction of the flow to report/collect data for flow-front positions as a function of time. The study adopts the FVM-VOF-based two-phase flow model together with an implicit time-stepping scheme, i.e., a dual-time formulation solution method with a preconditioned pseudo-time derivative. Initially, three time-step sizes, 5 s (small), 25 s, and 50 s (large), are evaluated to examine their impact on numerical saturation lines at various fabric porosities, 40%, 50%, and 60%, for a two-dimensional (2D) rectangular mould, and predictions are then compared with the well-known analytical Darcy. This is followed by a three-dimensional (3D) curved mould for a fillet L-shaped structure, wherein the degree-of-curvature of fibre preforms is incorporated using a User-Defined Function (UDF) to tailor the impregnation process. The developed approach shows its validation (1–5.7%) with theoretical calculations and experimental data for 2D and 3D cases, respectively. The results also stress that a shorter computational time can be achieved with a large time-step size while maintaining the same level of accuracy.

Keywords: liquid composite moulding; flow visualisation; numerical modelling; Volume of Fluid (VOF); multi-phase model

Citation: Alotaibi, H.; Abeykoon, C.; Soutis, C.; Jabbari, M. Numerical Simulation of Two-Phase Flow in Liquid Composite Moulding Using VOF-Based Implicit Time-Stepping Scheme. *J. Compos. Sci.* **2022**, *6*, 330. https://doi.org/10.3390/jcs6110330

Academic Editor: Jinyang Xu

Received: 23 September 2022
Accepted: 1 November 2022
Published: 3 November 2022

1. Introduction

During liquid composite moulding (LCM) processes, a liquid resin is injected/infused at a constant pressure/flow rate to impregnate a fibrous reinforcement [1,2]. This serves to fully saturate the placed/laid fibre preforms on the mould. This means that the fibre preforms are dry (unsaturated) during the pre-filling stage (c.f., Figure 1). The dry fibrous reinforcement needs to be processed with a thermosetting resin in order to produce a composite part. This process stage is referred to as the mould-filling stage, in which the injected/infused resin flow starts to impregnate the fibre preforms. The quality of the processed fibre–reinforcement composites would depend on this crucial process stage. This is due to the fact that gelation or early curing, as well as the emergence of voids, could occur within the impregnation process. To avoid these issues, the knowledge and control of filling time and flow-front advancement is required. For this reason, it is important to optimise the mould-filling process in an effective way to obtain good-quality composite parts/components. This optimisation can be done with the traditional trial and error method; however, this method shows drawbacks from the perspective of time and cost [3]. Recently, there have been alternative proposals to optimise such a process, and one of the most efficient methods is to use the available sophisticated numerical tools. This can include isothermal/non-isothermal conditions, but this study will focus on isothermal mould-filling processing.

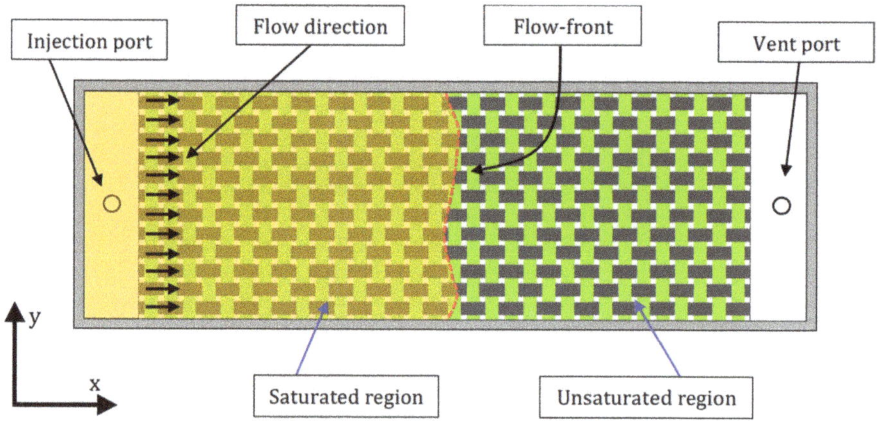

Figure 1. A schematic diagram illustrating mould-filling process parameters, saturation and unsaturation regions, and filling front advancement with uniform permeability.

Flow-front advancement has been discussed by a considerable amount of research [3–22]. This includes experimental [4–12] and numerical studies [5,8,13,22], in which the latter are subject to validation of the adopted numerical tool with the available experimental data or theoretical models. Experimental works for monitoring/tracking the flow progression used various types of techniques; these would include cameras, ultrasonic sensors, optic sensors, dielectric sensors, and pressure transducers. On the other hand, the developed numerical models involve a wide range of approaches that are based on the control volume finite element method (CVFEM), finite volume method (FVM), finite element method (FEM), and finite difference method (FDM). These discretisation methods allow the solving of multi-phase or multi-physics flow problems, and this could apply to resin flow in LCM processes. Hoes et al. [4] conducted an experimental work that involved filling front tracking at a constant pressure injection. The authors placed electric sensors together with pressure transducers in the bottom flat mould plate, wherein these sensors were located on straight lines and predefined in the connected data collection computer. This method enabled the automatic observation of flow-front positions as a function of time. Binetruy et al. [5] investigated, experimentally and theoretically, the interaction between micro- and macro-flow through unidirectional woven fabrics, and their impacts on the flow-front progression during the impregnation phase. The authors used a flat Plexiglass (transparent) piece on the top of the mould to allow visual observations of the filling process. The study highlighted that, for unsaturated (dry) tows, a lag of the intra-tow flow progression would lead the macro-flows to cause transverse fluid penetration into the yarns. This study concluded that the micropores could affect flow-front behaviour and slow the filling process. A contribution by Luce et al. [6] was presented by performing mould-filling experiments for multi-ply fibre preforms that consisted of different fabric architectures, such as 3-D woven and random mat (RM). Two cameras were placed on the top and side-view of the mould to monitor and characterise the in-plane and through-thickness advancing flow behaviour. The study showed that the flow-front advancement through RM behaves uniformly with rapid progression. However, this was different for the 3-D woven ply, in which the flow encountered progress resistance caused by the transverse tows, besides the emergence (presence) of out-plane flows resulting from advancing macro-infiltration. The application of ultrasonics was employed by Schmachtenberg et al. [7] to obtain the on-line monitoring of the flow-front propagation in the RTM manufacturing process. The generated experimental data stressed that signals, a transmission time, and an amplitude could explain the arriving flow-front, and hence achieve the sensitive optimisation of the RTM filling process. The work by Pierpaolo et al. [8] embedded dielectric sensors to capture pressure data of the infused unidirectional flow in LCM, as well as monitoring the progressing flow front. The results were validated against numerical modelling, with satisfaction in terms of pressure profile and flow progression. Their study incorporated multiscale modelling, in which a microstructure of the used fabric (2-D woven), was scanned to obtain a meso-scale image, whereby the meso-permeability was identified and introduced in the macro-scale simulation to validate the approach with the experimental work. At present, a broad range of sophisticated software packages are available to model mould-filling processes in LCM, including RTM and VARTM (see Table 1). These would include, but are not limited to, LIMS [13], PAM-RTM [14,15], OpenFoam [15], and COMSOL [16]. Simacek et al. [13] implemented an algorithmic code based on the finite element/control volume (FE/CV) method using the simulation tool LIMS to model tow saturation during the mould-filling process. The numerical study considered the dual-scale flow, and emphasised that the developed model is able to capture the filling progression as well as the required time to fully impregnate and saturated the fibre preforms. They also added that this approach can be applied to arbitrary complex shapes. Voller et al. [17] followed CVFEM, but using Fortran to track flow-front positions as a function of time. The work analysed the filling front with structured/unstructured mesh grids for fibre–reinforcement mats. The results showed good agreement with the well-known analytical Darcy for various time-step sizes. Multiscale modelling of bi-axial fabrics was developed by Tan et al. [3] to predict the filling front during RTM processes. The numerical model adopted the FE/CV approach to simulate a coupled macro- and micro-flow problem. The PoreFlow program was used in the study, which is based on the Fortran modular package, and hence showed its capability of simulating resin impregnation for dual-scale porous media. This was validated with experimental data obtained by the same authors. Grössing et al. [15] performed a numerical mould-filling analysis via two different simulation tools, i.e., OpenFoam and PAM-RTM. The study evaluated both tools and highlighted that non-porous and porous zones can be both modelled by OpenFoam, whereas this was different with Darcy-based PAM-RTM, whereupon porous zones can only be modelled. The authors also argued that the issue with PAM-RTM could be resolved by assigning non-porous regions as race-tracking zones with %100 porosity value. The numerical study was based on an RTM experimental work that involved two types of fibre preforms, such as NCF and UD. Both numerical results agreed well with the flow-front experimental data.

Table 1. Selection of numerical contributions that evaluated simulation tools for flow-front modelling.

References	Fabric Architecture	Injection Method	Flow Modelling	Computational Approach
Tan et al. [3]	Bi-axial	Unidirectional	Dual-scale	FE/CV-PoreFlow
Simacek et al. [13]	UD	Unidirectional	Dual-scale	FE/CV-LIMS
Oliveira et al. [14]	Fibre mats	Unidirectional	–	Darcy-based PAM-RTM
Grossing et al. [15]	UD/Triaxial NCF	Radial	Dual-scale	FVM-VOF OpenFoam
Sas et al. [16]	UD	Unidirectional	–	FEM-LSM COMSOL
Wei et al. [22]	Fibre mats	Unidirectional	–	FVM-VOF Moldex3D

The present work is motivated to contribute a simple, accurate technique using ANSYS Fluent to model resin flow advancement in the RTM/VARTM processes. The FVM-VOF-based multiphase flow approach is adopted to monitor the mould-filling process and to report flow-front positions as a function of time. This approach allows the prediction of the required time to fully saturate/impregnate the fibre preforms in simple and complex shapes. During the mould-filling simulation, the saturated, partially saturated, and unsaturated regions can be observed and located. The numerical simulations fit well with the analytical predictions and the experimental data for a rectilinear/channel flow injection.

2. Numerical Simulation Approach

2.1. Volume of Fluid (VOF)

The free-surface flows can be modelled by the prominent technique VOF, whereby the two immiscible fluids' interface position is tracked and located simultaneously. This VOF model can be solved either using implicit or explicit time formulation (see Section 2.2) throughout structure/unstructured fixed meshes—a Eulerian-based approach. Such formulations (interpolation schemes) will allow discretisation of the volume fraction (tracking of the interface) by applying widely adopted options (in ANSYS Fluent)—geometric reconstruction (a piecewise-linear approach) and Modified HRIC (High-Resolution Interface Capturing)—for explicit and implicit, respectively [23–25]. This is to obtain face fluxes (fluid advection) for all filled and empty cells, and also near-interface (partially filled) cells [23–25]. Momentum and continuity equations are applied to each cell throughout the domain, in which the flow motion is solved by Navier–Stokes equations (N–S) based on the FVM discretisation (control-volume-based) approach. For RTM/VARTM filling simulation, the volume fraction in each control volume (computational grid cell) denotes either one of the phases (resin or air), or a mixture of both phases (partial saturation). This approach follows a transient, laminar, incompressible, multiphase flow problem with Newtonian behaviour and an isothermal filling condition. Thus, the continuity/transport equation can be written as follows:

$$\frac{\partial s_i}{\partial t} + \nabla \cdot (s_i \mathbf{u}) = 0 \tag{1}$$

where $\mathbf{u}$ is the volume-averaged velocity, t is the time, and s_i denotes a phase volume fraction such that s_f is the fluid (e.g., resin), while s_a defines air (or empty) regions. This form assumes a constant fluid density and applies to fixed/stationary control volumes in this study. s_f varies from 0 to 1 in each grid cell, and this would indicate full saturation for cells with a value of 1, partial saturation for cells with a value ranging from $0 < s_f < 1$, and unsaturation for cells with a value of 0. Since the resin flow impregnates a porous material, the momentum equation used by the simulation tool incorporates a source term to allow porous media modelling. Equation (2) demonstrates the momentum equation after applying the above-mentioned assumptions, while Equation (3) expresses the source/sink term:

$$\frac{\partial}{\partial t}(\rho \mathbf{u}) + \nabla(\rho \mathbf{u}\mathbf{u}) = -\nabla p + \mu \nabla^2 \mathbf{u} + S \tag{2}$$

$$S = -\frac{\mu}{\mathbf{K}}\mathbf{u} \tag{3}$$

Here, ∇p is the pressure gradient, ρ is the fluid density, μ is the fluid viscosity, S is the sink term, and $\mathbf{K}$ is the permeability tensor. This filling modelling is performed on a rectangular mould (2-D RTM) in a macro-scale flow problem. The porous medium permeability is isotropic; therefore, the in-plane permeability can be denoted as $K_{xx} = K_{yy} = K$. The permeability value was computed in a previous work by the current authors at a dual-scale flow (inter- and intra-tow flow) through a plain-weave fabric unit cell at various aggregate porosity values: 40%, 50%, and 60%. Due to the fact that the unit cell can be taken as a representative volume element (RVE) of the woven fabric ply, the obtained in-plane dual-scale permeability is inputted in the source term to enable true viscous resistance of the woven model at a macroscopic level. The developed approach used the implicit scheme (see Section 2.2) provided by ANSYS Fluent, owing to the fact this method is preferred for slow flow movement problems, i.e., creeping flows, quasi-static motion, and static/dynamic motion [26]. In this method, the time-step size is independent of the results, and can be large to obtain a shorter computational time [17,26]. This work developed a flow-front tracking technique by creating points on a straight line (flow direction) at certain locations. Each point will report resin flow volume fraction data as a function of time; this would be similar to real-time monitoring RTM/VARTM experiments wherein sensors are embedded/placed (in fabrics or on mould).

As part of this study, three-dimensional flow simulations of complex shapes are included to examine the flow-front prediction model for a real filling process of a composite component. An experimental work by Geng et al. [27] was selected, wherein a curved mould was used with a VARTM process type. Hence, the present work added a set of equations (see Equation (4)) that accounts for the effect of curvature on resin flow through the curved regions during the resin impregnation of fibre preforms. This is done by using the User-Defined Functions (UDF) to interpret/compile these equations with a C-language-based code.

$$D_c = 2\sin^{-1}\left(\frac{C}{2R_c}\right), \quad \theta = \frac{D_c}{2} \tag{4a}$$

$$K_H(\theta) = K_H\cos(\theta) \tag{4b}$$

$$K_V(\theta) = K_V\sin(\theta) \tag{4c}$$

The degree of curvature and the deflection angle are D_c and θ, respectively. Here, R_c is the radius of curvature, and C is the cord length. Since part of the filling process is vertical, the permeability is set as a function of curvature for both vertical and horizontal resin flow progressions.

2.2. Time-Stepping Scheme

A temporal discretisation in transient simulations is defined by space and time. This means that an integration is required for each term in the differential equations with every time-step Δt. To apply such a time discretisation, there are two prominent schemes, i.e., implicit and explicit. These methods are to be carefully selected, since a diverse range of restrictions arise in each scheme in certain circumstances. The implicit method is applicable to slow motions (e.g., creeping flows), and it can handle long-duration analyses that vary from seconds to days [26]. On the other hand, the explicit case is suitable for rapid motions (e.g., drop tests, shocks, etc.), and it is restricted by the Courant–Friedrich–Lewy (CFL) condition, in which each grid cell uses the same time-step that depends on the courant number, cell volume, and the sum of outcoming fluxes [26]. Therefore, this work finds the implicit time-stepping a suitable technique for such a flow problem, and an evaluation was conducted to examine the impact of time-stepping size on the flow-front modelling. It is noteworthy that implicit time-stepping is a dual-time formulation that adopts a preconditioned pseudo-time (inner iterations) derivative at each physical time-step (Euler backward) to provide an accurate transient solution [26,28]. Figure 2 describes the time discretisation methods and how the time-dependent equations are computed differently throughout the domain. This shows that a function of a quantity/variable $f(\varphi)$ is evaluated at a future time level $(n+1)$ for the implicit case, while $f(\varphi)$ is computed at the current time level (n) for the explicit time-stepping method. The general expression used in the current numerical simulation is demonstrated below.

$$\frac{\varphi^{n+1} - \varphi^{n}}{\Delta t} = f(\varphi) \tag{5a}$$

$$\begin{cases} n+1 = t + \Delta t \\ n = t \\ n-1 = t - \Delta t \end{cases} \tag{5b}$$

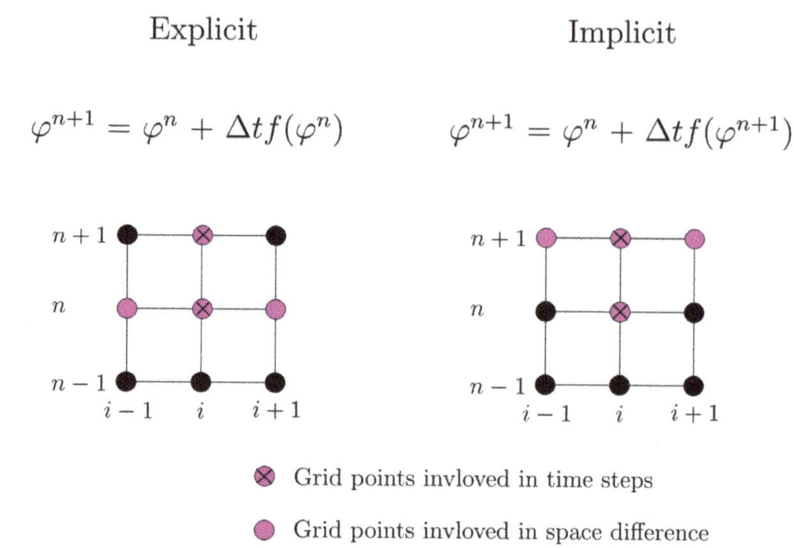

Figure 2. Unsteady flow solution: temporal discretisation methods and time-stepping schemes.

2.3. *Darcy's Law for a Transient Flow*

The general form of Darcy's law (Equation (6)) for an isothermal filling flow problem can be integrated and simplified to suit transient flow conditions [29]. At a constant pressure injection for one-directional flow movement (e.g., x-coordinate), Equation (6) can be rewritten as shown in Equation (7). By further simplification and integration of Equation (7), the position of the flow-front can be obtained by Equation (8).

$$\mathbf{u} = -\frac{\mathbf{K}}{\mu}\nabla p \quad (6)$$

$$\frac{u_x}{\phi_o} = \frac{dx}{dt} = -\frac{K}{\mu\phi_o}\frac{\Delta p}{\Delta x} \quad (7)$$

$$x_f = \sqrt{\frac{2Kp_o}{\mu\phi_o}t_f} \quad (8)$$

where x_f is the flow-front position, t_f is the flow-front time, ϕ_o is the porosity of the medium, and since the pressure at the flow-front can be assumed as zero, the pressure difference Δp becomes equal to the injection pressure (p_o) [29]. Thus, the proposed approach in the present study can be subjected to a comparison analysis with the above-mentioned theoretical solution for validation. On this premise, the developed numerical model will be also feasible for complex structural shapes, and this would contribute to the knowledge and decision of LCM optimisation. This is in terms of the optimal cycle time and location of injection ports/vents, thereby achieving void-free and good-quality composite parts. Figure 3 illustrates the geometry and boundary conditions used in the numerical analysis for the 2-D rectangular mould filling involving the flow-front tracking points.

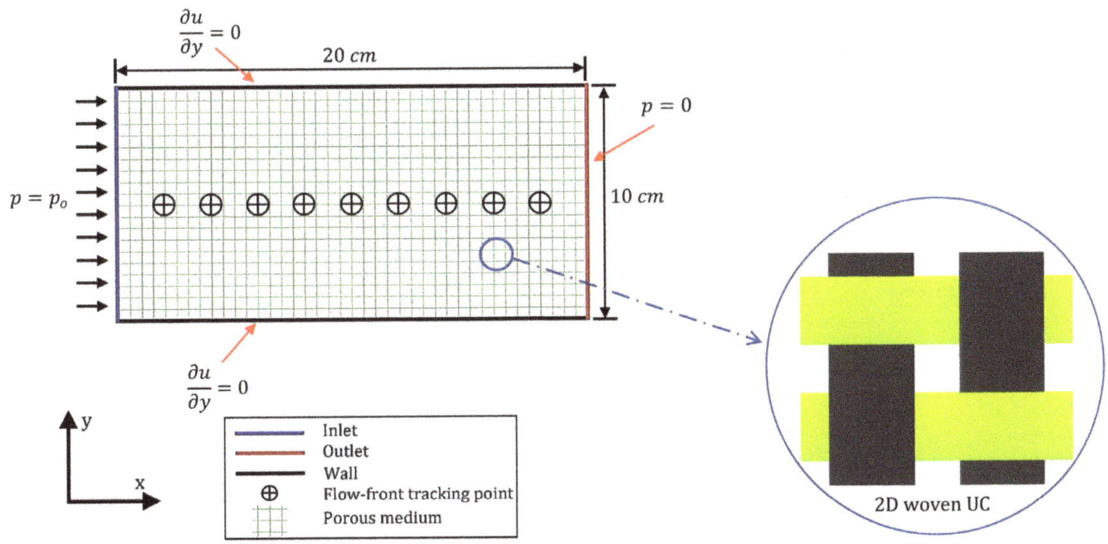

Figure 3. Geometry and boundary conditions used in the numerical simulation, including flow-front tracking points.

3. Results and Discussion

3.1. Two-Dimensional Rectangular Mould for Regular Shapes

The numerical simulations were performed at various fibre preform aggregate porosities (ϕ_o) 40%, 50%, and 60%, with in-plane dual-scale permeabilities ($K[\text{m}^2]$) 4.89×10^{-11}, 9.45×10^{-11}, and 2.09×10^{-10}, respectively. Filling front analysis was computed at a constant pressure injection ($p_o = 10$ [kPa]), and with fluid flow properties $\rho = 1300$ [kg/m^3], and $\mu = 0.15$ [Pa · s]. Three time step-sizes, 5 s, 25 s, and 50 s, were selected based on the generated grids/meshes for the transient multiphase flow through a 2-D macro-scale geometry. These different step sizes yielded significant factors in terms of the actual flow-front position or the so-called saturation lines and the computational time. It appeared that the partially filled control volumes (i.e., partially saturated region) could indicate the actual flow-front for all three time-step sizes. This can also be seen in similar numerical studies, such as [3], in which the flow-front was observed within the partially filled computational grid cells. Accordingly, the results stressed the fact that the flow-front progression with 40–50% resin volume fraction agreed well with the analytical solutions for all cases. Less computational time was observed at large time-steps such as 50 s; however, smaller step sizes are preferred (e.g., determined by grid cell size divided by fluid velocity value) [26]. By applying this, sharp flow-front behaviour could be seen in which the partial saturation zone was reduced (c.f., Figure 4). Consequently, filling front positions would be more manifested or perceivable, and an accurate on-line flow control could be attained. It should be noted that different implicit-scheme-based time-steps would offer the same required time for the mould-filling process; nonetheless, computational time processing would vary. For instance, to accomplish a fully saturated fibre preform with the geometry dimensions and boundary conditions illustrated in Figure 5 for the 50% porosity value case, 1600*s* is needed. This was achieved by all time-stepping sizes, 5*s*, 25*s*, and 50*s*, but with different numbers of time-steps: 320, 64, and 32 respectively. This means that a shorter computational time can be obtained with the maximum time-stepping size, while maintaining the same accuracy of the output results, and this can be explained by Figure 5 together with Table 2. Therefore, when supposing that the user is interested in a shorter computational time to optimise the mould-filling process, it would be recommended to follow the maximum time-stepping in accordance with the mesh/grid element size. It is worth mentioning that an Intel Core i7-1165G7 Processor (Central Processing Unit CPU) was used to run the present numerical simulations. Figure 5 presents the computed results for various woven fabric aggregate porosities, 40%, 50%, and 60%, with different time-step sizes, 5s, 25s, and 50s, for each porous medium case. The results showed perfect alignment with ones calculated by transient Darcy for all flow-front simulations. Therefore, the numerical approach proposed in this study shows its reliability and can be confidently used to optimise RTM/VARTM filling processes.

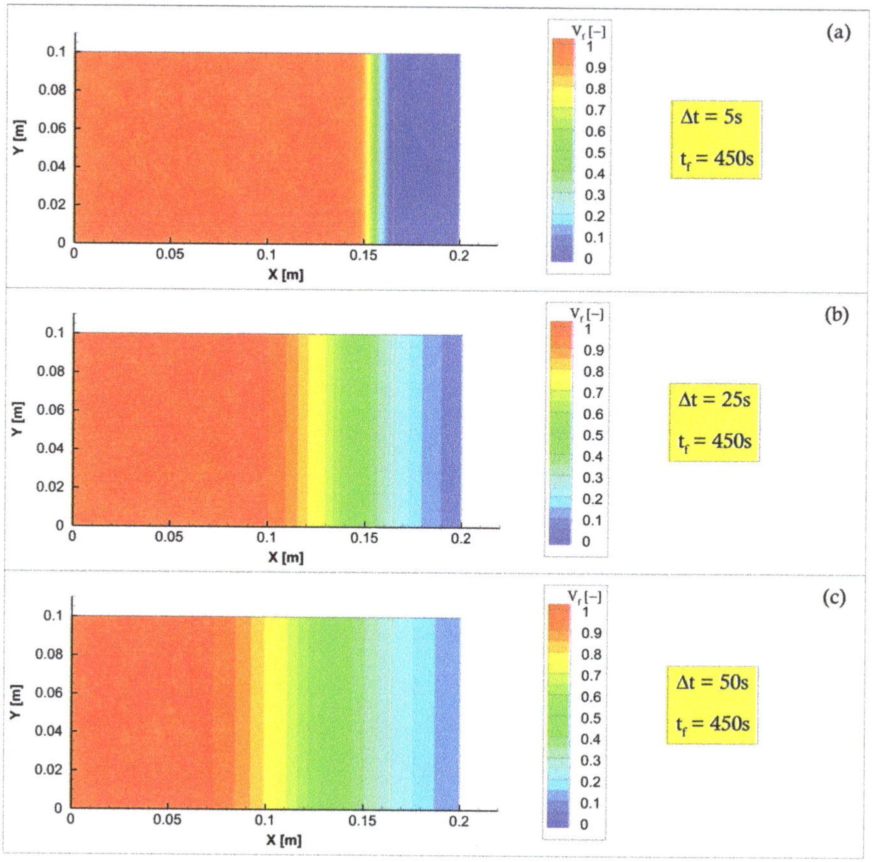

Figure 4. Numerical rectilinear/channel flow saturation throughout a porous medium with 50% aggregate porosity (ϕ_o) at a constant injection pressure for different time-step sizes: 5 s, 25 s, and 50 s. (**a**) 5 s, (**b**) 25 s, and (**c**) 50 s.

Table 2. Computational processing time with different time-stepping sizes for each medium porosity.

Medium Porosity [%]	Time-Stepping Size [s]	CPU [s]	Real Time [s]
40	5	19.75	481
	25	5.44	89
	50	2.97	57
50	5	14.67	319
	25	3.76	67
	50	2.2	40
60	5	8.14	148
	25	2.26	42
	50	1.55	26

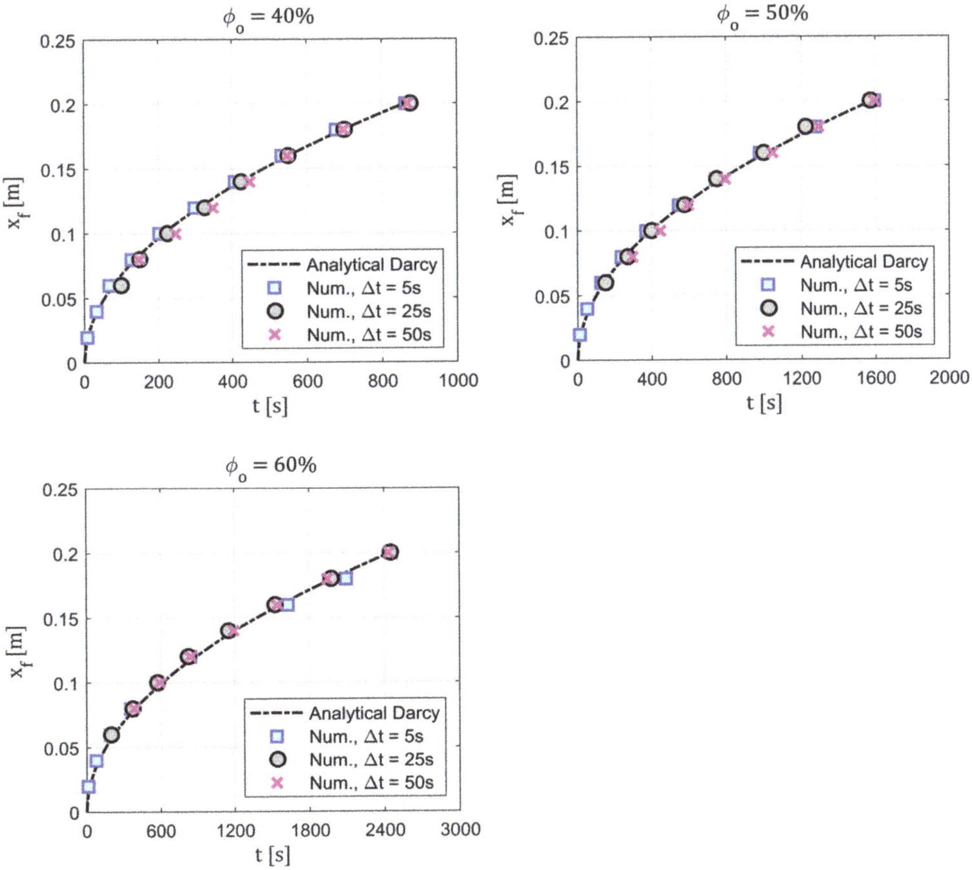

Figure 5. A comparative analysis for prediction of flow-front position as a function of time.

3.2. Three-Dimensional Curved Mould for Complex Shapes

For a complex shape, a study by Geng et al. [27] was selected to examine the current approach against the populated experimental data. Figure 6 shows a schematic diagram of the experimental setup that was used by [27], in which VARTM was adopted. Geng et al. [27] managed to perform experimental works for curved composite shapes using VARTM, and this was done with different bend angles, i.e., 180, 120, 90, and 60 degrees.

Therefore, this paper assesses the developed numerical flow-front prediction for a three dimensional complex shape during the mould-filling process. This is applied to curved non-crimp fabric (NCF) plates with 90 degree bending, and simulations are conducted on single and multiple layers based on the VARTM experiment by Geng et al. [27]. The permeabilities of fibre preforms are set to be 2.98×10^{-11} $[m^2]$ and 4.6×10^{-11} $[m^2]$, with fibre volume fractions (V_f) 22% and 40% for 1-layer and 6-layer, respectively [27]. This is along with the flow properties 968 $[kg/m^3]$ and 0.35 $[Pa \cdot s]$ for density and viscosity and a volume-average velocity range $(\mathbf{u})$ 0.00025–0.00035 $[m/s]$ [27]. With such a curved composite part, the curvature region is said to be impacting the resin impregnation, as was thoroughly discussed by Alotaibi et al. [30], and hence it is considered in the present work. In such a case, the resin flow will be affected by the degree of curvature as long as it progresses within the curved zone—see Figure 7. This would show 90 degrees of curvature (D_c) with a deflection angle impact range $(0° \leq \theta \leq 45°)$. Thereby, a set of equations is required to be incorporated with the flow equations—see Equation (7)—and this is done using the User-Defined Function (UDF) in ANSYS Fluent. The results give a good impression of the capability of the current numerical model, in which most of the tracking flow-front points (c.f., Figure 8) fit well, showing $\leq$ 5.7% discrepancy with the experimental data. This is in addition to the mould-filling time, wherein times of 760 s and 475 s were achieved for 1-ply and 6-ply, respectively, in comparison to 750 s and 470 s experimental fill time. Figure 9 portrays the resin flow progression contours/outlines for 1-ply and 6-ply NCFs, and with a more permeable fibre perform (e.g., 6-ply case), this would hasten the flow advancement. It should be noted that a hexahedron mesh topology is preferred when applying the relevant degree-of-curvature equations through the UDF, since it provides a sequential quadrilateral cell type, and this makes it compatible with the degree-of-curvature concept. The present numerical methodology proves its capability to monitor and predict resin flow advancement in RTM/VARTM processes that include regular and complex shapes, and it also offers a simple and accurate technique that permits efficient computational modelling.

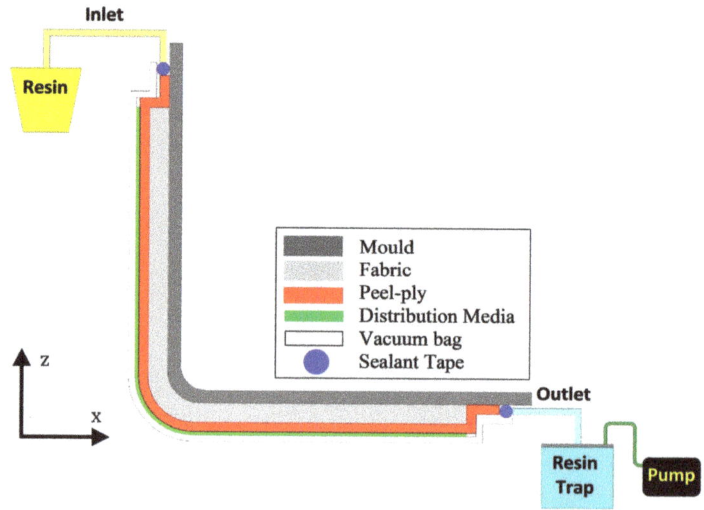

Figure 6. A VARTM experimental setup used for a curved or L-shaped composite components [27].

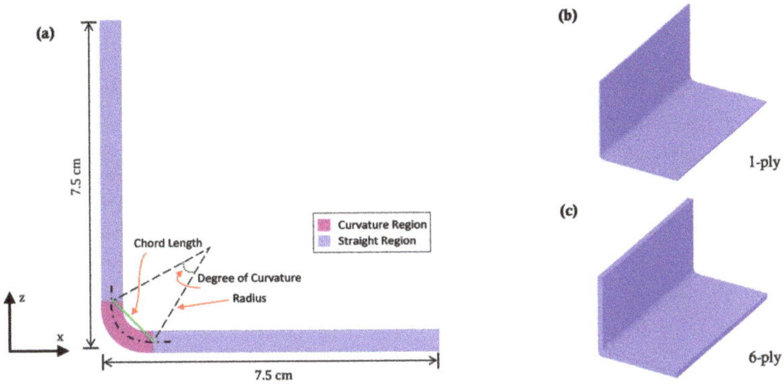

Figure 7. A schematic diagram of (**a**) curvature and straight regions of the L-shaped composite part, (**b**) 1-ply fibre preform with 1.46 mm thickness, and (**c**) 6-ply fibre preform with a 4.84 mm thickness.

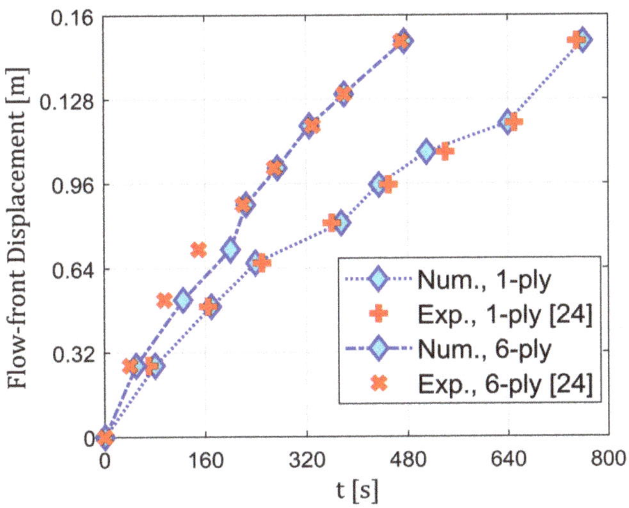

Figure 8. Numerical flow-front predictions vs. experimental observations for a complex shape with single and multiple plies.

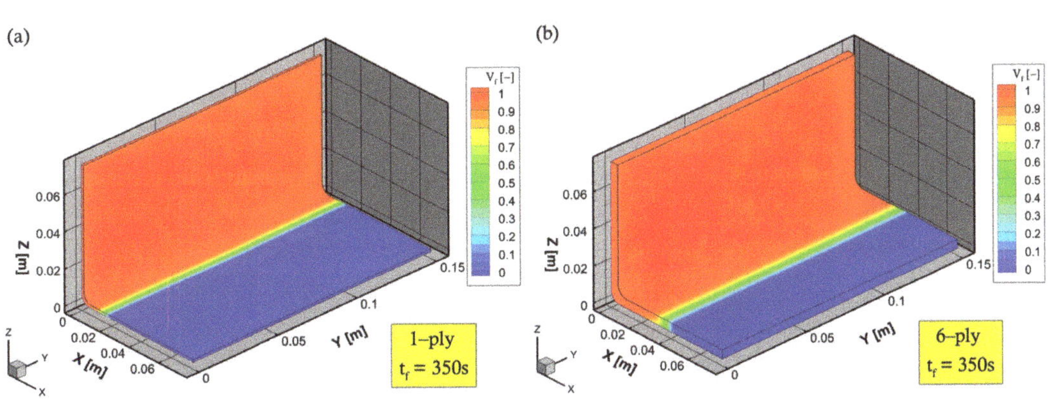

Figure 9. Three-dimensional numerical mould-filling process of a curved-type (L-shaped) composite component. (**a**) 1-ply (**b**) 6-ply.

4. Conclusions

A numerical technique was proposed using the data-track-point feature of ANSYS Fluent to predict the flow-front position as a function of time throughout the computational domain, as a means to simultaneously monitor filling progression during LCM processes (e.g., RTM/VARTM). The results stressed that the grid-independent solution (implicit VOF) of structured or unstructured control volumes does not impact the output of flow parameters, while allowing less computational time at larger time-step sizes. The simulation approach followed the FVM-VOF-based multiphase flow problem together with implicit time discretisation. The numerical model has been validated with Darcy's law and an experimental work [27] for a transient flow at various aggregate porosities of dual-scale fabrics. It is noteworthy that the present work assumes no chemical conversion of the fluid flow (i.e., a constant viscosity) during the impregnation process. A future work will include a cure-temperature-time-dependent viscosity to characterise their impacts on the impregnation/saturation of complex (woven) fabric structures with heterogeneous permeability.

Author Contributions: Conceptualisation, H.A. and M.J.; Investigation, H.A.; Software, H.A.; Writing—original draft, H.A.; Writing—review and editing, H.A., M.J., C.A. and C.S.; Supervision, M.J., C.A. and C.S. All authors have read and agreed to the published version of the manuscript.

Funding: This research received no external funding.

Institutional Review Board Statement: Not applicable.

Informed Consent Statement: Not applicable.

Data Availability Statement: Data are contained within the article.

Conflicts of Interest: The authors declare no conflict of interest.

References

1. Månson, J.A.E.; Wakeman, M.D.; Bernet, N. *Composite Processing and Manufacturing—An Overview*; Comprehensive Composite Materials; Elsevier: Amsterdam, The Netherlands, 2000; pp. 577–607; ISBN 978-0-08-042993-9.
2. Ermanni, P.; Di Fratta, C.; Trochu, F. Molding: Liquid composite molding(Lcm). In *Wiley Encyclopedia of Composites*; John Wiley & Sons, Inc.: Hoboken, NJ, USA, 2012; p. weoc153.
3. Tan, H.; Pillai, K.M. Multiscale modeling of unsaturated flow in dual-scale fiber preforms of liquid composite molding I: Isothermal flows. *Compos. Part A Appl. Sci.* **2012**, *43*, 14–28. [CrossRef]
4. Hoes, K.; Dinescu, D.; Sol, H.; Vanheule, M.; Parnas, R.S.; Luo, Y.; Verpoest, I. New set-up for measurement of permeability properties of fibrous reinforcements for RTM. *Compos. Part A Appl. Sci.* **2002**, *33*, 959–969. [CrossRef]
5. Binétruy, C.; Hilaire, B.; Pabiot, J. The interactions between flows occurring inside and outside fabric tows during rtm. *Compos. Sci. Technol.* **1997**, *57*, 587–596. [CrossRef]
6. Luce, T.L.; Advani, S.G.; Howard, J.G.; Parnas, R.S. Permeability characterization. Part 2: Flow behavior in multiple-layer preforms. *Polym. Compos.* **1995**, *16*, 446–458. [CrossRef]
7. Schmachtenberg, E.; Schulte zur Heide, J.; Töpker, J. Application of ultrasonics for the process control of Resin Transfer Moulding (Rtm). *Polym. Test.* **2005**, *24*, 330–338. [CrossRef]
8. Carlone, P.; Rubino, F.; Paradiso, V.; Tucci, F. Multi-scale modeling and online monitoring of resin flow through dual-scale textiles in liquid composite molding processes. *Int. J. Adv. Manuf. Technol.* **2018**, *96*, 2215–2230. [CrossRef]
9. Babu, B.Z.; Pillai, K.M. Experimental investigation of the effect of fiber-mat architecture on the unsaturated flow in liquid composite molding. *J. Compos. Mater.* **2004**, *38*, 57–79. [CrossRef]
10. Lawrence, J.M.; Barr, J.; Karmakar, R.; Advani, S.G. Characterization of preform permeability in the presence of race tracking. *Compos. Part A Appl. Sci.* **2004**, *35*, 1393–1405. [CrossRef]
11. Di Fratta, C.; Klunker, F.; Trochu, F.; Ermanni, P. Characterization of textile permeability as a function of fiber volume content with a single unidirectional injection experiment. *Compos. Part A Appl. Sci.* **2015**, *77*, 238–247. [CrossRef]
12. Di Fratta, C.; Koutsoukis, G.; Klunker, F.; Ermanni, P. Fast method to monitor the flow front and control injection parameters in resin transfer molding using pressure sensors. *J. Compos. Mater.* **2016**, *50*, 2941–2957. [CrossRef]
13. Simacek, P.; Advani, S.G. A numerical model to predict fiber tow saturation during liquid composite molding. *Compos. Sci. Technol.* **2003**, *63*, 1725–1736. [CrossRef]
14. Rodrigues, I.; Amico, S.C.; Souza, J.A.; de Lima, A.G.B. Numerical analysis of the resin transfer molding process via pam-rtm software. *Defect Diffus. Forum* **2015**, *365*, 88–93.

15. Grössing, H.; Stadlmajer, N.; Fauster, E.; Fleischmann, M.; Schledjewski, R. Flow front advancement during composite processing: Predictions from numerical filling simulation tools in comparison with real-world experiments. *Polym. Compos.* **2016**, *37*, 2782–2793. [CrossRef]
16. Şaş, H.S. Modeling of Particle Filled Resin Impregnation in Compression Resin Transfer Molding. Master's Thesis, Middle East Technical University, Ankara, Turkey, 2010.
17. Voller, V.R.; Peng, S. An algorithm for analysis of polymer filling of molds. *Polym. Eng. Sci.* **1995**, *35*, 1758–1765. [CrossRef]
18. Facciotto, S.; Simacek, P.; Advani, S.G.; Middendorf, P. Modeling of anisotropic dual scale flow in RTM using the finite elements method. *Compos. Part B Eng.* **2021**, *214*, 108735. [CrossRef]
19. Nielsen, D.R.; Pitchumani, R. Closed-loop flow control in resin transfer molding using real-time numerical process simulations. *Compos. Sci. Technol.* **2002**, *62*, 283–298. [CrossRef]
20. Lam, Y.C.; Joshi, S.C.; Liu, X.L. Numerical simulation of the mould-filling process in resin-transfer moulding. *Compos. Sci. Technol.* **2000**, *60*, 845–855. [CrossRef]
21. Matsuzaki, R.; Kobayashi, S.; Todoroki, A.; Mizutani, Y. Flow control by progressive forecasting using numerical simulation during vacuum-assisted resin transfer molding. *Compos. Part A Appl. Sci.* **2013**, *45*, 79–87. [CrossRef]
22. Wei, B.J.; Chuang, Y.C.; Wang, K.H.; Yao, Y. Model-assisted control of flow front in resin transfer molding based on real-time estimation of permeability/porosity ratio. *Polymers* **2016**, *8*, 337. [CrossRef]
23. Young, J.B. An equation of state for steam for turbomachinery and other flow calculations. *J. Eng. Gas Turbines Power* **1988**, *110*, 1–7. [CrossRef]
24. Nguyen, V.T.; Vu, D.T.; Park, W.G.; Jung, Y.R. Numerical analysis of water impact forces using a dual-time pseudo-compressibility method and volume-of-fluid interface tracking algorithm. *Comput. Fluids* **2014**, *103*, 18–33. [CrossRef]
25. Nguyen, V.T.; Park, W.G. A free surface flow solver for complex three-dimensional water impact problems based on the VOF method: A Free Surface Flow Solver for Complex 3D Water Impact Problems. *Int. J. Numer. Methods Fluids* **2016**, *82*, 3–34. [CrossRef]
26. ANSYS. *ANSYS Academic Research Fluent, Release 19.2, Help System, Theory Guide*; ANSYS, Inc.: Canonsburg, PA, USA, 2019.
27. Geng, Y.; Jiang, J.; Chen, N. Local impregnation behavior and simulation of non-crimp fabric on curved plates in vacuum assisted resin transfer molding. *Compos. Struct.* **2019**, *208*, 517–524. [CrossRef]
28. Venkateswaran, S.; Merkle, C. Dual time-stepping and preconditioning for unsteady computations. In Proceedings of the 33rd Aerospace Sciences Meeting and Exhibit, Reno, NV, USA, 9–12 January 1995; American Institute of Aeronautics and Astronautics: Reno, NV, USA, 1995.
29. Cai, Z.. Analysis of mold filling in rtm process. *J. Compos. Mater.* **1992**, *26*, 1310–1338.
30. Alotaibi, H.; Jabbari, M.; Soutis, C. A numerical analysis of resin flow in woven fabrics: Effect of local tow curvature on dual-scale permeability. *Materials* **2021**, *14*, 405. [CrossRef] [PubMed]

Journal of
Composites Science

MDPI

Article

Filling Time Reduction in Liquid Composite Molding Processes

Felice Rubino [1,*], Fausto Tucci [2], Vitantonio Esperto [2] and Pierpaolo Carlone [2]

[1] DIMME–Durability and Mechanical Integrity of Structural Materials, Escuela Superior de Ciencias Experimentales y Tecnología, Department of Chemical Universidad Rey Juan Carlos, C/Tulipán, 28933 Móstoles, Spain
[2] Department of Industrial Engineering, University of Salerno, Via Giovanni Paolo II, 132, 84084 Fisciano, Italy
* Correspondence: felice.rubino@urjc.es

Abstract: The quality of Liquid Composite Molding (LCM) manufactured components is strictly related to the fibrous preform impregnation. As Darcy's law suggests, the resin flow is influenced by the pressure gradient, geometrical features of the reinforcement, and resin viscosity. The former two parameters are dictated by the requirements of the component and other constraints; therefore, they are hardly modifiable during the process. Resin preheating increases its fluency, thus enhancing the impregnation and saturation flow, and reducing the mold filling time. In the present work, a microwave heating system has been integrated within a vacuum bag resin infusion process, to analyze the effect of the online preheating on the fiber impregnation. To monitor the resin flow a dielectric sensors-based system is used. Results from resin infusion tests conducted with and without the resin pre-heating were compared: the outcomes indicated an advance of approximately 190 s of the flow front when microwave heating is applied with respect to the unheated tests.

Keywords: liquid composite molding; microwave preheating; dielectric flow monitoring

Citation: Rubino, F.; Tucci, F.; Esperto, V.; Carlone, P. Filling Time Reduction in Liquid Composite Molding Processes. *J. Compos. Sci.* **2022**, *6*, 222. https://doi.org/10.3390/jcs6080222

Academic Editor: Jinyang Xu

Received: 12 July 2022
Accepted: 2 August 2022
Published: 4 August 2022

1. Introduction

Liquid composite molding (LCM) processes, such as resin transfer molding (RTM) or Seemann composites resin infusion molding (SCRIMP) processes have been addressed by composite industries as a promising technology to manufacture polymeric matrix composites out-of-autoclave. Some of these processes are particularly interesting for the industry involved in the production of large-scale structures, even with complex shapes [1]. However, large scale diffusion of components depends on the possibility to lower the overall costs of the products and scale-up the technology to a mass production, always guaranteeing the quality of the manufactured composite structures [2].

In LCM processes, the final quality of the products is strictly connected to the impregnation and curing phases [3]. Dry spots or excess in resin, delamination or cracks, and residual stress are some of the most common flaws that can occur during the manufacturing compromising the performance and the integrity of the structure if the mentioned aspects are not carefully designed and monitored [4,5]. Impregnation defects can be related to the incompatibility of the two main phases involved. From this point of view, the binder plays a key role [6].

Sensing techniques have been developed to monitor both resin flow front progression and the cure degree in thermoset matrix composite manufacturing processes. They include, but are not limited to, the use of optical FBG sensors [7–9] pressure transducers [10], thermocouples [11–14], SMART weave sensor [15,16], electrical time domain reflectometry (ETDR) [15], ultrasonic, dielectric and piezoelectric sensors [17–22]. In their previous works, authors developed a sensing system based on dielectric analysis (DEA) to monitor the resin flow progression during the Resin Infusion process [19,20]. DEA relies on the measurement of the dielectric properties, i.e., permittivity and ionic conductivity of the test material: dielectric material is placed between electrodes forming a capacitor and an alternating electric field is applied across the plates, then permittivity, loss factor and ionic conductivity can be determined from the output current [23]. Dielectric sensors, consisting of parallel plates placed on both sides of a mold, were implemented in a lab-scale LCM apparatus and provide pieces of information on the resin arrival at the sensor position by detecting variation in the dielectric properties of the medium (glass fibers plus resin) [20,24]. Since the electrodes can be placed outside of the composite laminate, the sensing system is less invasive and does not affect either the integrity of the manufactured parts or the surface finishing of the part.

In addition, the parallel plate dielectric sensors are characterized by a higher scanning depth, which makes this design suitable also for thick composite [24,25]. The plane plates dielectric sensors require that the sensing areas of the electrodes must be parallel to each other, and the reciprocal distance must be known and kept constant during the entire infusion process. This limited the usage of these types of dielectric sensors only to specific classes of LCM processes using rigid molds, such as RTM, which ensures that these requirements are observed. LCM processes involving flexible upper mold, such as Vacuum Assisted Resin Infusion (VARI) or SCRIMP processes, were coupled only with co-planar dielectric sensors [16,26]. To the authors' best knowledge, no attempts have yet been made to apply parallel-plates dielectric sensors to monitor these processes.

In addition to the monitoring and the control of impregnation and curing phases to ensure the quality of the composite parts, the scale-up of the composite industry to a mass-production is also limited by the manufacturing time of a component, dictated by the time required to fully impregnate the fibrous reinforcement and the time to complete the curing of the resin. This is especially true in the case of big components, such as boat hulls (usually manufactured by the SCRIMP process), where the impregnation is particularly long and the curing of resin is deliberatively kept slow to avoid gelation prior to the complete wetting [27]. In this scenario, enhancing strategies of the resin flow through the fabric is of paramount relevance to achieve a uniform impregnation of the fibers and optimize the filling time, mitigating the manufacturing flaws and contributing to a reduction in the overall production time of a composite structure. Among the strategies investigated, the reduction in resin viscosity by means of mold temperature increases [28]; preform or resin preheating [28–31] can improve the flow through the preform and, thus, decrease the impregnation time. Dealing with thermosetting resin systems involves time-temperature constraints related to its reactivity and the consequent reduction in pot life. From this point of view, microwave-based heating systems guarantee a high efficiency thermal energy transfer. Reductions in filling time of about 13% and 25% have been obtained in the case of non-reactive and reactive systems, respectively, by optimizing the microwave heating apparatus [32–35].

Previous experiments have been dedicated to the study of the resin flow through a dry fibrous preform sealed between rigid molds [32–35]. In the present article, the author investigated the application of microwave preheating to the SCRIMP process to reduce the filling time. Parallel plate dielectric sensors were also implemented to monitor the resin flow. An ad-hoc system was developed to install the electrodes on the vacuum bag and ensure the correct alignment of the plates during the whole process.

2. Materials and Methods

The liquid composite molding (LCM) experimental was conducted based on the following preliminary requisite materials along with the ancillary materials. HexForce E-glass twill 2/2 fabric was used as reinforcing material. 12 layers of the glass fiber fabric were cut into rectangular plies with dimensions 300 mm × 240 mm. Main properties of the reinforcement are indicated in Table 1.

Table 1. Reinforcement properties.

Reinforcement	Glass Fiber (E-Glass)
Weave style for intended preform	Twill 2/2 fabric
Areal density	390 g/m^2
Fabric thickness	0.3 mm
Construction	90° warp/weft

Epoxy resin SX 10 was utilized upon the room temperature premixing with the Epoxy-based hardener in the mixing ratio of 100:26. The resin properties are mentioned in Table 2. Its rheological behavior measured by rheometry testing is depicted in Figure 1 and described by the following equation:

$$\eta = A_\eta \, exp\left(\frac{B_\eta}{R\,T} + C_\eta \alpha\right), \tag{1}$$

where the viscosity η depends on the pre-exponential term $A_\eta = 7.093 \times 10^{-8}$ [Pa·s], the calibration coefficients $B_\eta = 3.999 \left[\mathrm{J{\cdot}mol^{-1}}\right]$ and $C_\eta = 1.63$, the universal constant of gases $R = 8.314 \left[\mathrm{J\,K^{-1}{\cdot}mol^{-1}}\right]$, the temperature T expressed in Kelvin degrees, and the degree of cure of the resin system α.

Table 2. Resin matrix properties.

Matrix	Epoxy SX10 EVO
Viscosity at 25 °C	0.250~1.20 Pa·s
Gardner index	3
Density at 20 °C	1100 kg/m^3
Flash point	>100 °C

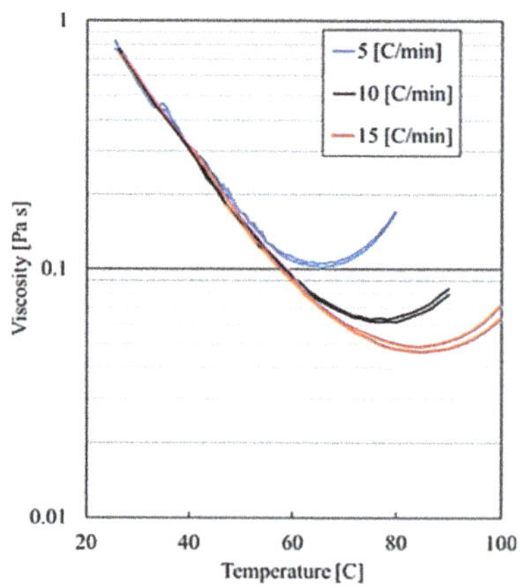

Figure 1. Viscosity of resin system as function of temperature and heating rate.

The vacuum bagging setup is shown in Figures 2 and 3, and it consists in the following steps:

1. Application of PVA release agent on the upper side of the mold;
2. Positioning of 12 sheets of glass fiber;
3. Positioning of peel ply and distribution media;
4. Positioning of the vacuum bag.

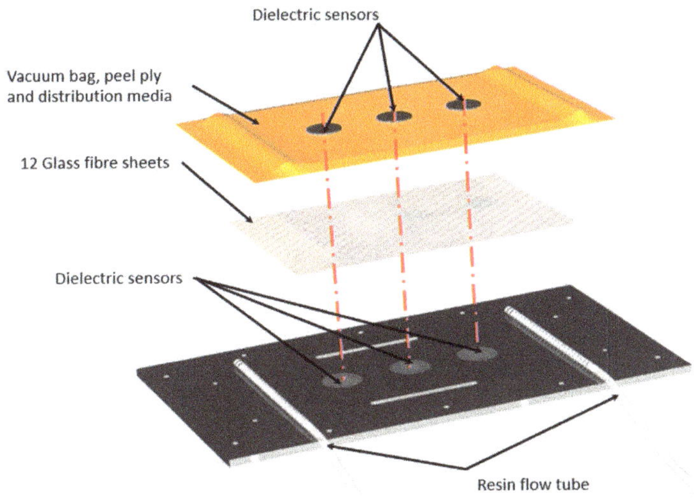

Figure 2. Vacuum bagging setup scheme.

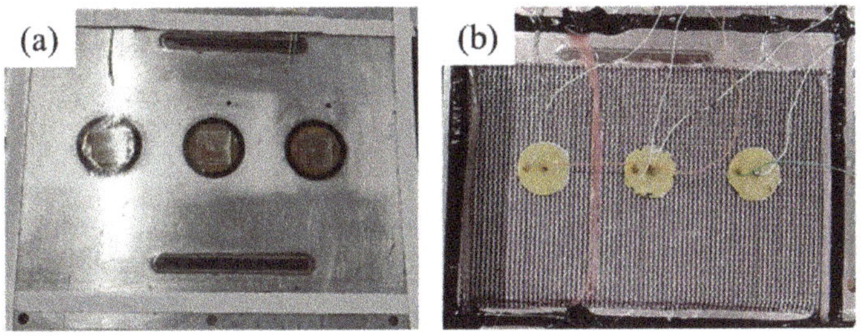

Figure 3. (**a**) Lower half-mold embedding the dielectric armatures. (**b**) Upper armatures mounted along with the soldered connecting wires over the vacuum bag during the resin flow progression.

The used ancillary materials, such as resin flow tubes and sealant tapes, were able to withstand high temperature. This selection was taken considering the temperature increase due to microwave preheating.

The lower armatures of the dielectric sensors are embedded in the rigid mold, as illustrated in Figures 2 and 3a. The armatures consist of square copper plates (25 × 25 mm^2) located at 60, 150, and 240 mm from the resin inlet, respectively. The three upper armatures of the capacitive sensors were fixed on the vacuum bag in correspondence to the lower ones. The eyelets milled in the rigid lower and sealed by transparent plastic allow to visually monitor the bottom-side flow. Two cameras were focused on the vacuum bag and on the eyelet during the entire test to monitor the position of the resin flow front.

Spiral wraps were inserted for the easy entry and exit of the resin under the vacuum bag. The vacuum bag encompassing the fibrous preform was sealed using the sealant tape avoiding any spot for external air insertion. The vacuum was induced by attaching the resin outlet tube to the vacuum pump. The resin inlet was clamped and negative air pressure of 0.9 bar was maintained to place the entire arrangement under vacuum conditions. Figure 4 shows the two schemes adopted to carry out the laboratory tests: the upper scheme represents the setup without microwave preheating; the lower is the setup involving the microwave preheating.

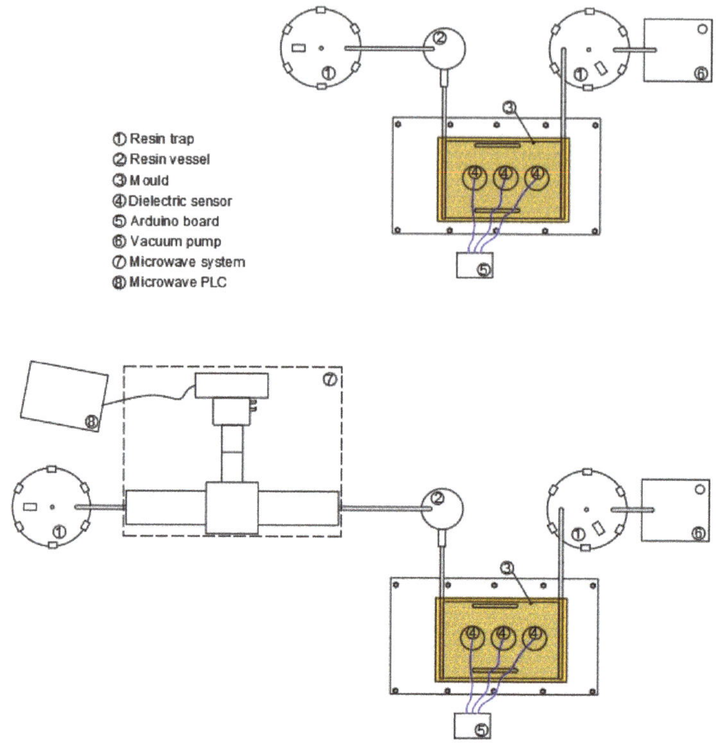

Figure 4. Schematic representation of the experimental setup of the resin infusion mold equipped with the dielectric and thermal acquiring system.

The microwave facility, depicted in Figure 4, consists of a 2 kW microwave generator, stainless steel waveguides, cylindrical resonance cavity, and ancillary tools. The choice of a microwave-based system to preheat the resin is based on the thermo-chemical and rheological behavior of the thermosets. Indeed, these polymeric systems are characterized by short pot-lives, which are further reduced when increasing their temperature. This entails the necessity for an efficient volumetric heating system. More details about the design and optimization of the apparatus can be found in previous articles [32–34]. An intermediate vessel has been placed between the exit of the microwave cavity and the resin inlet into the vacuum bag. The vessel works as a buffer to decouple the heating system and the LCM apparatus to avoid a potential mismatch between the resin flow rate and the amount of energy provided by the microwaves [34], which could lead to an overheating of the catalyzed resin. The resin is driven through the microwave preheating cavity to the buffer vessel by positive pressure. The resin flow through the resonant cavity and the microwave power emitted have been calibrated by performing preliminary heating tests. A resin flow of 0.38l/min was set, while the power emitted by the magnetron was 2 kW.

3. Results and Discussion

Dielectric measurements and visual analysis of the flow front performed during the two tests are reported in Figures 5 and 6.

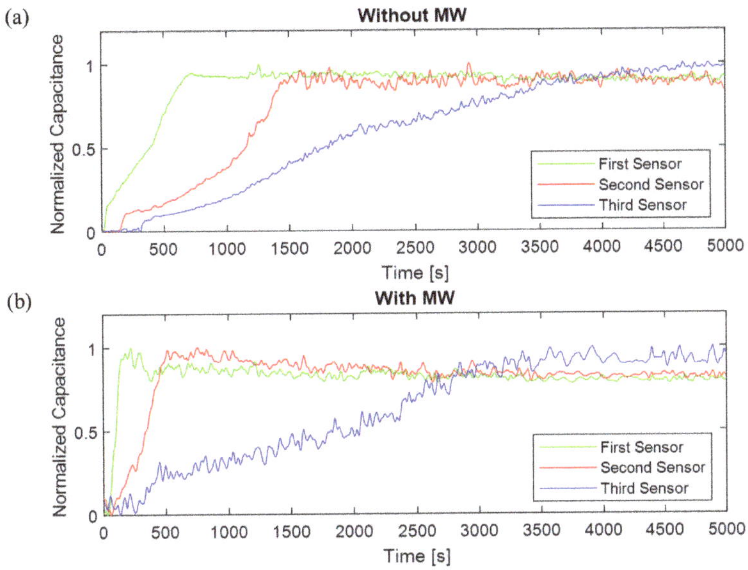

Figure 5. Dielectric acquisition in conventional resin infusion (**a**) and in resin infusion test with microwave preheating (**b**).

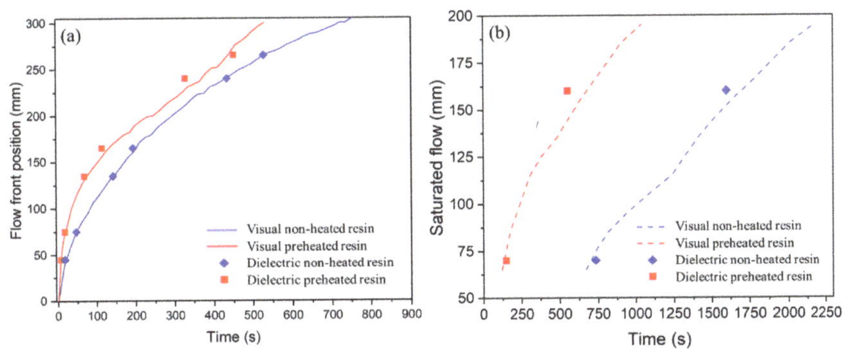

Figure 6. Visual resin flow front profiles compared with dielectric signals for resin arrival and preform saturation acquired during the two tests from top side (**a**) and bottom side (**b**).

The dielectric sensors detect variations in the capacitance of medium contained between the two armatures as the resin flow reach the sensor locations [19,20]. The observable variations in the signals during the infusion can be ascribed to the resin flow through the glass fiber fabric. Three distinct trends in the capacitance curves can be detected, which are more evident in the test without preheating where the resin flow is slower than that in the tests with resin preheating. The signal shows an initial increasing step, ranging from 8 to 15% of the saturated signal and is related to the flow of the resin through the flow media: the resin, indeed, promptly impregnated the distribution web covering almost immediately the sensing area due to its high permeability, which can be two orders of magnitude higher that the textile or even more. After that, the resin progressively fills the glass fabric layer below the flow media and the capacitance profile proceeds with a reduced slope until it reaches a plateau. At this point the proportion between resin and fibers stabilizes and the signal remains almost constant or without significant variations. Fluctuations of the signals are due to the continuous flow of the resin through the preform in the sensing areas. The capacitance curves of the three sensors present different slope for each phase of the infusion and are due to the reduction in the resin velocity as expected in case of applied constant gradient pressure, typical of the SCRIMP process. Therefore, the impregnation is faster at the beginning of the infusion and slows down as the infusion proceeds (it is also visible in the flow front profiles described in Figure 6)

By comparing the profiles of the signals from the two experiments, the effect of the preheating on the behavior of the resin is visible. The signal detected by sensor 1, located at 60 mm from the resin inlet, reaches the plateau with a remarkably advance when compared to the non-heated resin case (Figure 5). Despite that the first step related to the resin flow through the flow media, which does not show remarkable differences, the steeper curve describing the second stage of the infusion evidences how the impregnation of the glass fabric layers proceeds faster in the test with preheated resin reaching the saturation in approximately ten seconds after the arrival of the resin at the sensing location.

A similar trend can be observed at the locations of sensors 2 and 3, where saturation of the preform with preheated resin was achieved in less than half the time required by the non-heated resin. It is worthy of noting that the effect of the resin preheating cannot be appreciated at the very beginning of the infusion when resin first goes through the flow medium: the high permeability has a predominant effect on the resin velocity than the reduction in viscosity from the temperature increasing [36].

The advancement profiles of the resin flow front captured by the cameras during the two experiments are reported in Figure 6. The two profiles in Figure 6a refer to the resin advancing on the flow media, while Figure 6b shows the flow front of the resin acquired on the mold surface from the eyelet (Figure 2). The position of the flow front has been also acquired from the dielectric signals and reported in both graphs for the tests with unheated and preheated resin.

In both tests, the resin flow velocity decreases during the infusion. Indeed, the resin initially advances pushed by a high-pressure gradient and, during the infusion, the gradient decreases in relation to the advancement of the resin describing the conventional flow through a porous medium [36]. Figure 6 shows the difference between the two test cases. Indeed, the microwave preheated resin flow front reaches the vent in less than 70% of the non-heated resin. In the earliest 100 mm the microwave preheated resin flow front is more than twice faster if compared to the non-heated case: 2.3 mm/s for the preheated resin, 1.0 mm/s for the conventional process. The velocity difference decreases as the process continues. In the last 100 mm, the average flow front velocities are 0.34 mm/s and 0.23 mm/s, respectively.

The analysis of the flow-front detections acquired from the eyelets of the bottom half-mold indicate the complete impregnation of the preform, which is delayed with respect to the top flow due to the difference in the permeability of flow media and fiber fabric. Due to the design of the rigid half mold, the bottom flow front can be acquired only on the eyelet, which ranges from 65 to 195 mm from the inlet. Therefore, only data from sensors 1 and 2 can be correlated to the visual analysis, since the sensing area of sensor 3 is not covered by the eyelet. From data reported in Figure 6, it is possible to observe that while the reduction in filling time on the top of the preform is approximately 25% at 195 mm from the inlet, the difference in the bottom flow between the tests with preheating and non-heated resin is around 50%. Clearly, the flow through the distribution medium is less affected by the preheating, thanks to the high permeability of that medium (it is visible also in the dielectric signals, as mentioned before); on the other hand, it has a significant influence on the flow through the fabric where the reduction in the viscosity plays a major role in facilitating the impregnation of the fibers. The beneficial effect of microwave preheating, as reported in previous works [32–34], is related to the rheological behavior of thermoset resins: the temperature reached by the resin at the exit of the microwave cavity was approximately 36 °C, while the room temperature at which the test with non-heated resin was conducted was 22 °C; at 36 °C the viscosity decreases from 0.8 Pa·s up to 0.4 Pa·s, as shown in Figure 1. Nevertheless, the thermal energy conferred to the resin must be carefully tuned to avoid premature gelation of the thermoset. At the operative temperature reached by the resin in the test with preheating, no gelation occurred for the time required to complete the impregnation of the preform.

Table 3 summarizes the times of the resin flow when it reaches the sensing locations; the arrival on the sensor edges determining the first reaction of the sensors and the time when the resin reached the opposite edge and fully cover the sensor. The saturation time corresponds to the stabilization of the dielectric signals (Figure 5) and it represents the moment when the resin reaches the end of the sensing area on the bottom of the mold, and it has fully impregnated the whole thickness of the preform. The graphs in Figure 6 show the good agreement between the visual and dielectric analyses and the reliability of the latter in detecting the resin flow on both the top and the bottom of the mold.

Table 3. Times of resin flow arrival at sensing locations for tests with non-heated and preheated resin acquired by dielectric sensors and the saturation of the dielectric signals (in the parenthesis are reported the distances from the inlet port).

	Non-Heated Resin Infusion	Preheated Resin Infusion
	Time [s]	
Arrive to S1 (50 mm)	17.0	6.0
Arrive to end of S1 (70 mm)	46.0	16.5
Saturation S1	730	145
Arrive to S2 (140 mm)	140	66.5
Arrive to end of S2 (160 mm)	190	115
Saturation S2	1600	550
Arrive to S3 (230 mm)	415	325
Arrive to end of S3 (250 mm)	485	450
Saturation S3	5020	3810

By analyzing the dielectric data, it is possible to appreciate the benefits of the microwave heating method: a remarkable reduction in the infusion time in the test with microwave preheated resin was registered with the shortening of the saturation times of the three sensors by approximately 80%, 65%, and 30%, respectively. The top flow registered a smaller decrease in times of approximately 64 %, 52%, and 7%, respectively, on the three sensors.

Figures 7 and 8 depict a qualitative representation of the flow front derived from the data of the dielectric sensors at specific moments of the infusion.

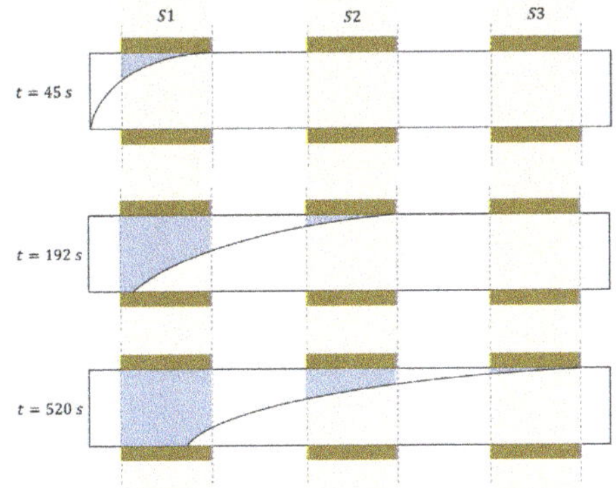

Figure 7. Schematic representation of flow front profile at the arrival of the resin on the sensing area for infusion with non-heated resin.

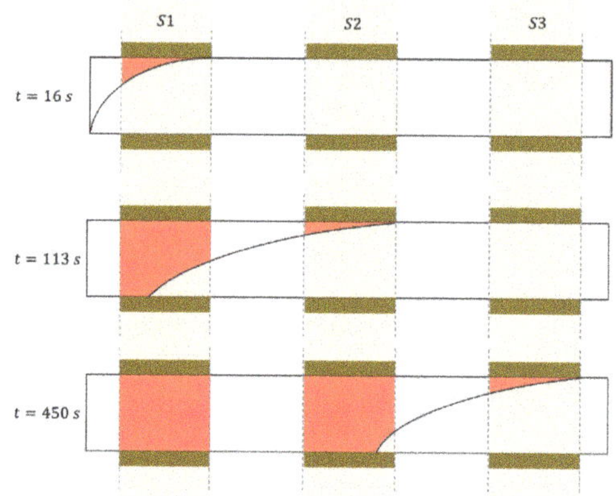

Figure 8. Schematic representation of flow front profile at the arrival of the resin on the sensing area for the infusion with microwave preheated resin.

In the SCRIMP process, the presence of a distribution medium determines the formation of two main flows: the first through the free-fibers region (i.e., the distribution medium) and the second through the fiber preform region (bulk porous medium). The former consists of a longitudinal flow from the inlet to the vent, while the second present a combination of a transverse flow (i.e., the out-of-plane flow through the thickness of the fiber preform) and a longitudinal one. This gives place to two regions characterized by different flow behaviors: a fully saturated zone, where the fluid flows through the preform with a velocity profile constant along the thickness of the preform, and a partially saturated zone, characterized by longitudinal and transverse flows through the thickness (Figure 8, time 450 s).

In the former region, the velocity vector is parallel to the main flow direction representing a fully developed flow, and meaning that the flow is substantially unidirectional with no significant crossed flows.

The flow in the latter zone is bi-directional (visual analysis indicated that in the in-plane flow no variations of the resin velocity occurred along the transversal direction) since the resin permeates from the distribution media in the through-the-thickness direction. The higher resin velocity along the longitudinal direction in the distribution medium determined a complex shape flow front. This feature characterizes the unsaturated region [36].

The formation of the unsaturated region, as mentioned before, occurred as the infusion begins due to the high longitudinal permeability of the distribution medium compared to the reinforcement one. In the unsaturated region, part of the liquid resin flows transversally from the distribution medium toward the preform, however, this transversal flow occurs to limited extent also in the saturated region determining the formation of a transition region between the saturated zone and flow front region [36].

Previous work by some of the authors [36] pointed out that the ratios between the thicknesses and between the permeabilities of the distribution medium and preform influence the extension of the unsaturated region and, hence, the delay between the first arrival of the resin to the vent and fully impregnation of the preform while other parameters, such as the fiber volume fraction or the compressibility of textiles, do not play a significant role. Therefore, in the present experimentation, the differences observed have to be ascribed to the change in rheological properties induced by the preheating being the other factors kept equal in the two tests.

In the infusion with non-heated resin, the unbalance in the permeability of distribution medium and fiber bulk determined a large difference in the resin velocity in the tow region resulting in a long unsaturated region (Figure 7). Indeed, when the resin reached the end of sensor 2 the preform in sensing area 1 is only partially filled, the signal is at 30% of its plateau and, by extension, it is possible assuming that the preform is impregnated by the same percentage. On the other hand, in the test with preheated resin the length of the unsaturated zone is far smaller and the preform in sensing area 1 is filled with the resin by almost 70%, consistently with the reduction in the time required by sensor 1 to reach the plateau in the two experiments (Table 3). Furthermore, it is possible to observe that when the resin reaches the end of sensor 3 the preform at sensor 1 location is still not fully impregnated: the resin reached that location almost 250 s in advance with respect to the saturation of the preform, which was approximately 77% (Table 3). The preform at the location of sensor 2, as consequence, result partially filled by only 17%. Conversely, in the test with preheating, the resin filled in a shorter time the preform thanks to the reduced viscosity and the enhanced flow resulting in a full impregnation at sensing area 1 and in a high level of saturation of the preform at the sensing area 2 of almost 80%. The analysis of the present preliminary experimentation indicates that microwave heating is effective to enhance the resin flow not only in the conventional RTM or VARTM processes [34] but also in the case of processes using a flexible half mold, such as SCRIMP, being able to further promote the impregnation and reduce the overall filling time more than that obtainable by using the sole distribution medium. The good agreement between dielectric signals and visual analysis also indicated the validity of the dielectric analysis and parallel-plate sensor for flexible-mold manufacturing processes. Further development involving the numerical analysis of the resin flow will be useful to strengthen the correlation between the actual position of the resin during the infusion and the signals from the dielectric sensors.

4. Conclusions

This paper compares the performances of the conventional resin infusion process and microwave preheated resin infusion in the case of flexible mold. The experiments conducted and the analysis of the achieved outcomes evidenced that the resin system preheating gives place to beneficial effects in the vacuum infusion processes, with a marked reduction in the cycle time. The reduction in the viscosity provokes an improvement of the flow, in agreement with Darcy's law. Considering the achieved results and what was discussed above, the following remarks can be drawn:

1. The unsaturated resin flow front is faster at the inlet due to the higher gradient of pressure. It decelerates during the infusion. In the present experiments, the final flow front velocity ranges between 15% and 22% of the initial velocity.
2. The flow front velocity is significantly affected by the microwave preheating: the resin reached vent 220 s earlier than in the non-heated resin case evidencing a reduction of 29% in the flow front crossing time.
3. The dielectric sensors detected a marked deceleration of the saturated resin flow as the distance from the resin inlet increased. The velocity decrease between sensor 1 (60 mm from the inlet) and sensor 3 (240 mm from the inlet) ranges between 84% and 92%.
4. The microwave preheating is beneficial to the infusion process. The dielectric measurement evidenced a decrease in overall saturation time of 50% correlated to a reduction in the length of the unsaturated region.

The presented results raise new interrogatives, such as the effects of microwave preheating on the fibrous preform saturation in vacuum bag infusion processes and resin system curing time, the effects related to the geometry and the architecture of the fibrous preform, or the influence of the binder applied, just to mention some of them. These aspects should be investigated by further analyses in future works.

Author Contributions: Conceptualization, F.R., F.T., V.E. and P.C.; methodology, F.R., F.T., V.E. and P.C.; formal analysis, F.R., F.T., V.E. and P.C.; investigation, F.R., F.T. and V.E.; resources, P.C.; data curation, F.R., F.T., V.E. and P.C.; writing—original draft preparation, F.R., F.T. and V.E.; writing—review and editing, F.R., F.T., V.E. and P.C.; visualization, F.R., F.T. and V.E.; supervision, P.C.; project administration, P.C.; funding acquisition, P.C. All authors have read and agreed to the published version of the manuscript.

Funding: This research received no external funding.

Institutional Review Board Statement: Not applicable.

Informed Consent Statement: Not applicable.

Data Availability Statement: Not applicable.

Conflicts of Interest: The authors declare no conflict of interest.

References

1. Boutier, S.; Fourel, J.; Sanial, P. How To Make a Choice Between Liquid Composite Molding Processes? In Proceedings of the 9th International Conference on Flow Processes in Composite Materials-FPCM-9, Montréal, QC, Canada, 8–10 July 2008; pp. 1–8.
2. Roth, Y.C.; Weinholdt, M.; Winkelmann, L. Liquid Composite Moulding-Enabler for the Automated Production of CFRP Aircfraft Components. In Proceedings of the ECCM 16 - 16th European Conference on Composite Materials, Seville, Spain, 22–26 June 2014.
3. Park, C.H.; Woo, L. Modeling Void Formation and Unsaturated Flow in Liquid Composite Molding Processes: A Survey and Review. *J. Reinf. Plast. Compos.* **2011**, *30*, 957–977. [CrossRef]
4. Kuentzer, N.; Simacek, P.; Advani, S.G.; Walsh, S. Correlation of Void Distribution to VARTM Manufacturing Techniques. *Compos. Part A Appl. Sci. Manuf.* **2007**, *38*, 802–813. [CrossRef]
5. Hamidi, Y.K.; Altan, M.C. Process Induced Defects in Liquid Molding Processes of Composites. *Int. Polym. Process.* **2017**, *32*, 527–544. [CrossRef]
6. Terekhov, I.V.; Chistyakov, E.M. Binders Used for the Manufacturing of Composite Materials by Liquid Composite Molding. *Polymers* **2022**, *14*, 87. [CrossRef] [PubMed]
7. Molimard, J.; Vacher, S.; Vautrin, A. Monitoring LCM Process by FBG Sensor Under Birefringence. *Strain* **2011**, *47*, 364–373. [CrossRef]
8. Gupta, N.; Sundaram, R. Fiber Optic Sensors for Monitoring Flow in Vacuum Enhanced Resin Infusion Technology (VERITy) Process. *Compos. Part A Appl. Sci. Manuf.* **2009**, *40*, 1065–1070. [CrossRef]
9. Tucci, F.; Rubino, F.; Carlone, P. Strain and Temperature Measurement in Pultrusion Processes by Fiber Bragg Grating Sensors. *AIP Conf. Proc.* **2018**, *1960*, 020036. [CrossRef]
10. Di Fratta, C.; Klunker, F.; Ermanni, P. A Methodology for Flow-Front Estimation in LCM Processes Based on Pressure Sensors. *Compos. Part A. Appl. Sci. Manuf.* **2013**, *47*, 1–11. [CrossRef]
11. Tuncol, G.; Danisman, M.; Kaynar, A.; Sozer, E.M. Constraints on Monitoring Resin Flow in the Resin Transfer Molding (RTM) Process by Using Thermocouple Sensors. *Compos. Part A. Appl. Sci. Manuf.* **2007**, *38*, 1363–1386. [CrossRef]
12. Tucci, F.; Bezerra, R.; Rubino, F.; Carlone, P. Multiphase Flow Simulation in Injection Pultrusion with Variable Properties. *Mater. Manuf. Process.* **2020**, *35*, 152–162. [CrossRef]
13. Vedernikov, A.; Safonov, A.; Tucci, F.; Carlone, P.; Akhatov, I. Modeling Spring-In of L-Shaped Structural Profiles Pultruded at Different Pulling Speeds. *Polymers* **2021**, *13*, 2748. [CrossRef] [PubMed]
14. Vedernikov, A.; Tucci, F.; Safonov, A.; Carlone, P.; Gusev, S.; Akhatov, I. Investigation on the Shape Distortions of Pultruded Profiles at Different Pulling Speed. *Procedia Manuf.* **2020**, *47*, 1–5. [CrossRef]
15. Dominauskas, A.; Heider, D.; Gillespie, J.W. Electric Time-Domain Reflectometry Sensor for Online Flow Sensing in Liquid Composite Molding Processing. *Compos. Part A Appl. Sci. Manuf.* **2003**, *34*, 67–74. [CrossRef]
16. Vaidya, U.K.; Jadhav, N.C.; Hosur, M.V.; Gillespie, J.W.; Fink, B.K. Assessment of Flow and Cure Monitoring Using Direct Current and Alternating Current Sensing in Vacuum-Assisted Resin Transfer Molding. *Smart Mater. Struct.* **2000**, *9*, 727–736. [CrossRef]
17. Hancioglu, M.; Sozer, E.M.; Advani, S.G. Comparison of In-Plane Resin Transfer Molding and Vacuum-Assisted Resin Transfer Molding 'Effective' Permeabilities Based on Mold Filling Experiments and Simulations. *J. Reinf. Plast. Compos.* **2020**, *39*, 31–44. [CrossRef]
18. Tifkitsis, K.I.; Skordos, A.A. A Novel Dielectric Sensor for Process Monitoring of Carbon Fibre Composites Manufacture. *Compos. Part A Appl. Sci. Manuf.* **2019**, *123*, 180–189. [CrossRef]
19. Carlone, P.; Rubino, F.; Paradiso, V.; Tucci, F. Multi-S.Scale Modeling and Online Monitoring of Resin Flow through Dual-Scale Textiles in Liquid Composite Molding Processes. *Int. J. Adv. Manuf. Technol.* **2018**, *96*, 2215–2230. [CrossRef]
20. Carlone, P.; Palazzo, G.S. Unsaturated and Saturated Flow Front Tracking in Liquid Composite Molding Processes Using Dielectric Sensors. *Appl. Compos. Mater.* **2015**, *22*, 543–557. [CrossRef]
21. Lee, H.L. *The Handbook of Dielectric Analysis and Cure Monitoring*, 2nd ed.; Lambient Technologies LLC: Cambridge, MA, USA, 2014.
22. Qing, X.; Liu, X.; Zhu, J.; Wang, Y. In-Situ Monitoring of Liquid Composite Molding Process Using Piezoelectric Sensor Network. *Struct. Health Monit.* **2020**. [CrossRef]

23. Abraham, D.; Mcilhagger, R. Glass Fibre Epoxy Composite Cure Monitoring Using Parallel Plate Dielectric Analysis in Comparison with Thermal and Mechanical Testing Techniques. *Compos. Part A. Appl. Sci. Manuf.* **1998**, *29*, 811–819. [CrossRef]
24. Hegg, M.C.; Ogale, A.; Mescher, A.; Mamishev, A.V.; Minaie, B. Remote Monitoring of Resin Transfer Molding Processes by Distributed Dielectric Sensors. *J. Compos. Mater.* **2005**, *39*, 1519–1539. [CrossRef]
25. McIlhagger, A.; Brown, D.; Hill, B. Development of a Dielectric System for the On-Line Cure Monitoring of the Resin Transfer Moulding Process. *Compos. Part A Appl. Sci. Manuf.* **2000**, *31*, 1373–1381. [CrossRef]
26. Kim, H.G.; Lee, D.G. Dielectric Cure Monitoring for Glass/Polyester Prepreg Composites. *Compos. Struct.* **2002**, *57*, 91–99. [CrossRef]
27. Rubino, F.; Nisticò, A.; Tucci, F.; Carlone, P. Marine Application of Fiber Reinforced Composites: A Review. *J. Mar. Sci. Eng.* **2020**, *8*, 26. [CrossRef]
28. Zhang, K.; Gu, Y.; Wang, S.; Li, M.; Zhang, Z. Effects of Preheating Temperature of the Mould on the Properties of Rapid-Curing Carbon Fibre Composites Fabricated by Vacuum-Assisted Resin Infusion Moulding. *Polym. Polym. Compos.* **2014**, *22*, 825–836. [CrossRef]
29. Zhang, K.M.; Gu, Y.Z.; Li, M.; Wang, S.K.; Zhang, Z.G. Resistive Heating of Carbon Fiber Aided Rapid Curing of Vacuum Assisted Resin Infusion Molding. *Adv. Mater. Res.* **2014**, *1030–1032*, 170–173. [CrossRef]
30. Johnson, M.S.; Rudd, C.D.; Hill, D.J. The Effect of Microwave Resin Preheating on the Quality of Laminates Produced by Resin Transfer Molding. *Polym. Compos.* **1997**, *18*, 185–197. [CrossRef]
31. Rudd, C.D.; Kendall, K.N. Towards a Manufacturing Technology for High-Volume Production of Composite Components. *Proc. Inst. Mech. Eng. Part B J. Eng. Manuf.* **1992**, *206*, 77–91. [CrossRef]
32. Esperto, V.; Durante, M.; Carlone, P.; Carrino, L. Resin Microwave Preheating in Liquid Composite Molding Process. *AIP Conf. Proc.* **2019**, *2113*, 110007. [CrossRef]
33. Rubino, F.; Paradiso, V.; Carlone, P. Flow Monitoring of Microwave Pre-Heated Resin in LCM Processes. *AIP Conf. Proc.* **2017**, *1896*, 030017. [CrossRef]
34. Rubino, F.; Esperto, V.; Tucci, F.; Carlone, P. Flow Enhancement in Liquid Composite Molding Processes by Online Microwave Resin Preheating. *Polym. Eng. Sci.* **2020**, *60*, 2377–2389. [CrossRef]
35. Esperto, V.; Boccarusso, L.; Durante, M.; Carrino, L.; Carlone, P. Permeability Analysis of Natural and Artificial Fiber Textiles for Liquid Composite Molding Process. *Procedia Manuf.* **2020**, *47*, 435–439. [CrossRef]
36. Rubino, F.; Carlone, P. A Semi-Analytical Model to Predict Infusion Time and Reinforcement Thickness in VARTM and SCRIMP Processes. *Polymers* **2019**, *11*, 20. [CrossRef] [PubMed]

MDPI
St. Alban-Anlage 66
4052 Basel
Switzerland
Tel. +41 61 683 77 34
Fax +41 61 302 89 18
www.mdpi.com

Journal of Composites Science Editorial Office
E-mail: jcs@mdpi.com
www.mdpi.com/journal/jcs

www.ingramcontent.com/pod-product-compliance
Lightning Source LLC
Chambersburg PA
CBHW061623210926
51850CB00009B/2418
* 9 7 8 3 0 3 6 5 8 1 2 0 0 *